MEMOIRS AND PROCEEDINGS

OF

THE MANCHESTER

LITERARY & PHILOSOPHICAL SOCIETY.

FOURTH SERIES.

SIXTH VOLUME.

That the quantity of heat capable of increasing the temperature of a lb of water (weighed in vacuo, and taken at between 55° and 60°) by one degree Fahr. requires for its evolution the expenditure of a mechanical force represented by the pressure of 772 lbs through the space of one foot.

J. P. Joule

MEMOIR

OF

JAMES PRESCOTT JOULE,

D.C.L. (OXON.), LL.D. (DUBL. ET EDIN.), F.R.S., HON. F.R.S.E.,
PRESIDENT OF THE LITERARY AND PHILOSOPHICAL SOCIETY
OF MANCHESTER,

F.C.S., DOC. NAT. PHIL. LUGD. BAT., SOC. PHIL. CANTAB.,
SOC. PHIL. GLASC., INST. MACH. ET NAUP, SCOT. ET SOC. ANTIQ. PERTH,
SOC. HONOR. INST. FR. (ACAD. SCI.),
CORRESP. SOC. REG. DAN. HAFN. TAURIN, BOLON., SOC. PHIL. NAT. BASIL.,
SOC. PHYS. FR. PAR. ET HAL.,
ET ACAD. AMER. SCI. ET ARTIB. ADSOC. HONOR.

BY

OSBORNE REYNOLDS, M.A., LL.D., F.R.S., Mem. Inst. C.E.,
Honorary Fellow of Queen's College, Cambridge,
Professor of Engineering, Owens College, Manchester.

MANCHESTER LITERARY AND PHILOSOPHICAL SOCIETY:
36, GEORGE STREET.
—
1892.

CONTENTS.

CHAPTER I.

                                                                             PAGE

INTRODUCTION.—The Mechanical Foundation of Physical Science.—'Matter, Living Force, and Heat.'—State of the Language and Knowledge of Physical Science in 1838.—No Recognized Measure of Mechanical Effect.—The Materiality of Heat.—Mechanical Origin of Heat.—Contrary Evidence of the Condensing Steam Engine.—Dependence of the Work Developed on Temperature.—Carnot's Law.—The Locomotive Obtrusive.—Suggestions as to Conversion of Heat into Work; Séguin, Mayer, and Colding.—Chemical and Physical Effects.—Discoveries of Oersted, Sturgeon, Ohm, and Faraday.—Invention of Electro-Magnetic Engine.—Evidence of Convertibility of Heat and Work not Recognized   1

CHAPTER II.

PARENTAGE AND EARLY LIFE.—Sees the First Train.—Education and Companionship with his Brother.—Association with Dalton.—Similarity of the Works of Dalton and Joule.—The Brothers' Vacations.—Their Amusements.—Under Treatment for the Spine.—Commencement of Life.—Continued Companionship.—Visits to Dalton.—Further Instruction in Chemistry. — Intercourse with Sturgeon and Members of this Society.—Joule a Dangerous Companion with a Gun. — British Association. — Dr. Scoresby.—Visit to Bradford.—Interruption of Companionhip.—Joule's activity as a Boy 25

CHAPTER III.

JOULE'S FIRST RESEARCH.—Starts to improve Sturgeon's Electro-Magnetic Engine.—Increases the Magnetic Force.—Does not realize His Problem.—Effects the Absolute Measure of 'Work.'—Finds that the Speed is Limited.—Seeks for the Limit in the Magnets.—Fails to find it.—Realizes the importance of Measuring the Current.—Constructs a Standard Galvanometer.—Repeats his Experiments.—Discovers Fundamental Law of Electro-Magnetic Attraction.—Contemplates Perpetual Motion.—Explains Law of Electro-Magnetic Attraction.—Measures Current, Velocity, Resistance, and estimates Zinc Consumed in Producing the Current.—Obtains 'Duty' per lb. of Zinc.—Realizes Resistance to the Current Induced by the Motion of the Magnet.—Refers to Faraday, Ohm, &c.—Determines Law of Induced Resistance.—Discovers Equivalence of Mechanical Effect to the Electric-Action, and Chemical-Action expended in its Production.—Introduces Absolute Electric Measurement. — Concludes that the Electro-Magnetic Engine can never compete with the Steam Engine.—Sees a great Philosophical Discovery before him. 33

CHAPTER IV.

SECOND RESEARCH.—Communicates Results to the Royal Society.—Joule's Motive.—Attributes Proportionality of Chemical and Mechanical Effects to their respective Quantitative Relations to the Electric Action.—Heat in Metallic Conductors.—Proportional to the Square of the Current. — Absolute Measures of Heat, Current, and Electromotive Force. — Heat Equivalent of Electrical Effect. — Heat Developed during Electrolysis.—Electric Origin of Heat.—Heats of Combustion and Electrolysis.—Intensity of Chemical Affinity of Combustibles.—Permanent and Transitory Voltaic Intensity.—Dependence of Affinity on Gaseous or Liquid States.—First Paper before the

CONTENTS. v

PAGE

Society.—The British Association.—Joint Research with Scoresby.—Heat Evolved during Electrolysis of Water.—Summary of Results.—Approaches Generalization.—Commences Third Research 45

CHAPTER V.

THIRD RESEARCH.—Heat Generated or Transferred?—Arrangement not Generation of Heat in Voltaic Apparatus.—The Heat Developed in the Entire Circuit by Magneto-Electricity not the Result of Arrangement.—Discovers Heat in the Revolving Armature.—Determines Relation between Heat and Electric-Action; the Same as with Voltaic Current.—Generation and Destruction of Heat by Mechanical Means by the Agency of Magneto-Electricity.—Constant Ratio between the Heat and the Power.—First Determination of the Mechanical Equivalent of Heat.—The Climax of Joule's Researches.—Attends British Association at Cork, 1843.—Dazzled by Possibility of Practical Results.—Conclusions not altogether Justified.—Postscript.—Mechanical Effect Converted into Heat by Friction.—Law of Conservation Realized 59

CHAPTER VI.

EFFORTS TO CONVINCE THE SCIENTIFIC WORLD.—Pre-eminence in Knowledge of Physical Science.—General Silence; the Highest Tribute to the Greatness of the Advance.—Friendly Sympathy.—Oakfield.—Researches in his New Laboratory.—Rarefaction and Condensation of Air.—Difficulties.—No Latent Heat.—Convertibility of Free Heat into "Work."—Dynamical Theory of Heat.—Development of Davy's Dynamical Theory of Gases.—New Theory of the Steam Engine.—Criticism of Carnot's Theory.—Heat Discharged into the Condenser.—Discussion of Results.—Joule ignores the Truth of Carnot's Theory.—Indestructibility of Caloric Proved.—Definite Dynamical Theory of Gases.—Imperfect acquaintance with Mechanical Philosophy.—Absolute Zero of Temperature.—First Determination with the Paddle.—Realization of Dynamical Significance of "Work."—Extent of Experimental Work.—Research with Scoresby ; Visits to Bradford.—Essay to the Institute of France.—Research on Effect of Magnetism on the Dimensions of Iron and Steel.—Joint Research with Sir Lyon Playfair; Atomic Volumes... 74

CHAPTER VII.

THE YEAR 1847.—Lecture at St. Ann's Church Reading Room.—Conservation of " Force."—Fresh Determination of Equivalent.—Verification of Laplace's Theory of Sound. —Joule's Paper accepted by the Institute of France.—Meeting of British Association at Oxford.—First Public Recognition of Joule's Discoveries.—Joule's Account.—Sir William Thomson's Account.—Marriage.—Shooting Stars. —The Adoption of Herapath's Hypothesis.—Determination of Velocities of Molecules of Gases and Theoretical Specific Heats 103

CHAPTER VIII.

JOULE'S VIEWS ACCEPTED BY THOMSON, RANKINE, AND CLAUSIUS.—Effect of Publication of Regnault's Researches. —Thomson's First Paper on Mechanical Effect by Thermal Agency.—Maintains Inconvertibility of Heat.—Note on Joule's Views.—Conversion of Heat into Work Denied.—Work into Heat Accepted.—Second Paper.—Greater Deference.—Accepts Joule's Difficulties as to Carnot's Axiom.—Thomson's Courage in Expressing his Doubts.—Discovery of Dissipation of Energy.—Rankine's Hypothetical Theory of Heat.—Acknowledges Joule's Hypothesis.—Hypothetical Foundation obscures General Laws.—Accepts Joule's Views.—Criticises Joule's Experiments.—Apology and Acceptance of Joule's Equivalent.—Joule suggests the form of Carnot's function.—Clausius Theory—Based on Joule's and Carnot's discoveries—Contains Hypotheses.—

CONTENTS. vii

Thomson's Third Paper—General Foundation without Hypotheses—Enthusiastic Acceptance of all Joule's Views.—Joule's Final Determination of the Mechanical Equivalent.—Historic Sketch.—Cordial Recognition of the Views and Work of his Predecessors and Contemporaries 115

CHAPTER IX.

MIDDLE LIFE.—Summer of his Life.—Acton Square.—Welcomes Mathematical Assistance.—Comparative Rest.—Amalgams.—Air Engine.—Joint Research with Thomson.—Vice-President of the Society.—Visit of Sir William Thomson.—Birth of his Daughter.—Royal Medal.—Death of his Wife.—Return to Oakfield.—Honours.—Electrical Welding.—Joint Research Resumed.—Memoir of Sturgeon.—Thermo-dynamical Properties of Solids.—Council of Royal Society.—Railway Accident.—Work at Oakfield.—Thorncliffe.—His Experiments Stopped.—President of the Society.—Honours.—Visits.—Small Scale Researches.—Determination of Mechanical Equivalent of Heat from the Thermal Effects of Electricity.—Propagation of Joule's Views.—Hirn's Verifications 135

CHAPTER X.

LATER LIFE.—Joule in 1869.—Estimation in the Society.—Sociability.—Copley Medal.—Institute of France.—Numerous Communications.—Display of Character.—Alleged Effect of Frost.—Performance of Electro-Magnetic Engines.—President of British Association.—First Failure of Health.—Verification of Final Determination of Mechanical Equivalent of Heat.—Change of Residence; 12, Wardle Road, Sale.—Royal Pension.—Honours.—Collected Scientific Papers by the Physical Society of London.—Declines to be President of British Association, 1887.—Failing Health.—Death.—Memorial Statue in Manchester.—Memorial Tablet in Westminster Abbey.—International Memorial.—Portraits of Joule 154

CONTENTS.

PAGE

APPENDIX TO PAGE 18.

DEVELOPMENT OF KNOWLEDGE OF HEAT FROM 1650.—Hooke's Vibratory Theory of Heat and Light.—Heat as a Measurable Fluid.—Heat as an Indestructible Fluid.—"Caloric" Invented by Lavoisier.—Extracts from "Micrographia," by Hooke.—Extracts from "Traité Eleméntaire de Chimie," Lavoisier 173

NOTE *A* TO PAGE 88.

REFERENCE TO THE VIEWS OF ROGET AND FARADAY.—"Force," destructible and "Force," indestructible.—Extract from Faraday's "Experimental Researches in Electricity"... 187

ERRATA.

Second paragraph, *for the full realizing of this general significance*, read *for the full realization of their general significance*.

p. 73, for *1893* read *1843*.

p. 110, fifth line from the bottom of the page, for *In June*, read *On August 18*.

MEMOIR
OF
JAMES PRESCOTT JOULE,
1891.

CHAPTER I.

INTRODUCTION.—*The Mechanical Foundation of Physical Science.*—'*Matter, Living Force, and Heat.*'—*State of the Language and Knowledge of Physical Science in 1838.* — *No Recognised Measure of Mechanical Effect.*—*The Materiality of Heat.*—*Mechanical Origin of Heat.*—*Contrary Evidence of the Condensing Steam Engine.*—*Dependence of the Work Developed on Temperature.* — *Carnot's Law.*—*The Locomotive Obtrusive.*—*Suggestions as to Conversion of Heat into Work*; *Séquin, Mayer, and Colding.*—*Chemical and Physical Effects.*—*Discoveries of Oersted, Sturgeon, Ohm, and Faraday.*—*Invention of Electro - Magnetic Engine.* — *Evidence of Convertibility of Heat and Work not Recognised.*

The three laws, the law of conservation of momentum, discovered by Newton, the law of conservation of the chemical elements, discovered by Dalton, and the law of the conservation of energy, discovered by Joule, constitute a complete mechanical foundation for physical science. The discoverer of a law is he who first generalizes whether he has or has not taken part in the discovery of the facts on which the generalization is made. Newton, Dalton and

Joule each of them, made discoveries which alone rendered generalization possible. And this was more particularly the case with Joule's discovery of the law of conservation of energy. This law, too, by completing the foundation of physical science rendered possible the grand generalization by which all branches of science are united in mechanical philosophy; so that it is to the memory of Joule that mankind owes its gratitude for the grandest generalization in the Universe—the complete mechanical foundation of physical science.

As a prelude to the biography of one whose work constituted his life, to a degree far beyond what is usual even with scientific men, it seems fitting to introduce his own full and clear exposition of the objects and subjects of his work, given after he had been led by his discoveries to the full realizing of this general significance, but three years before any philosopher was bold enough to follow him; and which as being the first exposition of the law of the universal conservation of energy, has necessarily a place in this Memoir.

*" In our notion of matter two ideas are generally included, namely, those of *impenetrability* and *extension*. By the extension of matter we mean the space which it occupies; by its impenetrability we mean that two bodies cannot exist at the same time in the same place. Impenetrability and extension cannot with much propriety be reckoned among the *properties* of matter, but deserve rather to be called its *definitions*, because nothing that does

---

* "On Matter, Living Force and Heat." By J. P. Joule, Secretary of the Manchester Literary and Philosophical Society. A Lecture at St. Ann's Church Reading-Room. 1847.

not possess the two qualities bears the name of matter. If we conceive of impenetrability and extension we have the idea of matter, and of matter only.

"Matter is endowed with an exceedingly great variety of wonderful properties, some of which are common to all matter, while others are present variously, so as to constitute a difference between one body and another. Of the first of these classes, the attraction of gravitation is one of the most important. We observe its presence readily in all solid bodies, the component parts of which are, in the opinion of Majocci, held together by this force. If we break the body in pieces, and remove the separate pieces to a distance from each other, they will still be found to attract each other, though in a very slight degree, owing to the force being one which diminishes very rapidly as the bodies are removed further from one another. The larger the bodies are the more powerful is the force of attraction subsisting between them. Hence, although the force of attraction between small bodies can only be appreciated by the most delicate apparatus except in the case of contact, that which is occasioned by a body of immense magnitude, such as the earth, becomes very considerable. This attraction of bodies towards the earth constitutes what is called their *weight* or *gravity*, and is always exactly proportional to the quantity of matter. Hence, if any body be found to weigh 2 lb., while another only weighs 1 lb., the former will contain exactly twice as much matter as the latter; and this is the case, whatever the bulk of the bodies may be: 2 lb. weight of air contains exactly twice the quantity of matter that 1 lb. of lead does.

"Matter is sometimes endowed with other kinds of attraction besides the attraction of gravitation; sometimes also it possesses the faculty of *repulsion*, by which force the particles tend to separate further from each other. Wherever these forces exist, they do not supersede the attraction of gravitation. Thus the weight of a piece of iron or steel is in no way affected by imparting to it the magnetic virtue.

"Besides the force of gravitation, there is another very remarkable property displayed in an equal degree by every kind of matter—its perseverance in any condition, whether of rest or motion, in which it may have been placed. This faculty has received the name of *inertia*, signifying passiveness, or the inability of any thing to change its own state. It is in consequence of this property that a body at rest cannot be set in motion without the application of a certain amount of force to it, and also that when once the body has been set in motion it will never stop of itself, but continue to move straight forwards with a uniform velocity until acted upon by another force, which, if applied contrary to the direction of motion, will retard it, if in the same direction will accelerate it, and if sideways will cause it to move in a curved direction. In the case in which the force is applied contrary in direction, but equal in degree to that which set the body first in motion, it will be entirely deprived of motion whatever time may have elapsed since the first impulse, and to whatever distance the body may have travelled.

"From these facts it is obvious that the force expended in setting a body in motion is carried by the body itself, and exists with it and in it, throughout the whole course of its motion. This force possessed by moving bodies is termed by mechanical philosophers *vis viva*, or *living force*. The term may be deemed by some inappropriate, inasmuch as there is no life, properly speaking, in question; but it is *useful* in order to distinguish the moving force from that which is stationary in its character, as the force of gravity. When, therefore, in the subsequent parts of this lecture I employ the term *living force,* you will understand that I simply mean the force of bodies in motion. The living force of bodies is regulated by their weight and by the velocity of their motion. You will readily understand that if a body of a certain weight possess a certain quantity of living force, twice as much living force will be possessed by a body of twice the weight, provided both bodies move with equal velocity. But the law by which the *velocity* of a body regulates its living force is not so obvious. At first sight one would imagine that the living force would be simply proportional to the velocity, so that if a body moved twice as fast as another, it would have twice the impetus or living force. Such, however, is not the case; for if three bodies of equal weight move with the respective velocities of 1, 2, and 3 miles per hour, their living forces will be found to be proportional to those numbers multiplied by themselves, viz., to 1×1, 2×2, 3×3, or 1, 4, and 9, the squares of 1, 2, and 3. This remarkable law may be proved in several ways. A bullet fired from a gun at a certain velocity will pierce a block of wood to only one quarter of the depth it would if propelled at twice the velocity. Again, if a cannon-ball were found to fly at a certain velocity when propelled by a given charge of gunpowder, and it were required to load the cannon so as to propel the ball with twice that velocity, it would be found necessary to employ four times the weight of powder previously used. Thus, also, it will be found that a railway-train going at 70 miles per hour possesses 100 times the impetus, or living force, that it does when travelling at 7 miles per hour.

"A body may be endowed with living force in several ways. It may receive it by the impact of another body, Thus, if a perfectly elastic ball be made to strike another similar ball of equal weight at rest, the striking ball will communicate the whole of its living force to the ball struck, and, remaining at rest itself, will cause the other ball to move in the same direction and with the same velocity that it did itself before the collision. Here we see an instance of the facility with which living force may be transferred from one body to another. A body may also be endowed with living force by means of the action of gravitation upon it through a certain distance. If I hold a ball at a certain height and drop it, it will have acquired when it arrives at the ground a degree of living force proportional to its weight and the height from which it has fallen. We see, then, that living force may be produced by the action of gravity through a given distance or space. We may, therefore, say that the former is of equal value, or *equivalent*, to the latter. Hence, if I raise a weight of 1 lb. to the height of one foot, so that gravity may act on it through that distance, I shall communicate to it that which is of equal value or equivalent to a certain amount of living force; if I raise the weight to twice the height, I shall communicate to it the equivalent of twice the quantity of living force. Hence, also, when we compress a spring, we communicate to it the equivalent to a certain amount of living force; for in that case we produce molecular attraction between the particles of the spring through the distance they are forced asunder, which is strictly analogous to the production of the attraction of gravitation through a certain distance.

"You will at once perceive that the living force of which

we have been speaking is one of the most important qualities with which matter can be endowed, and, as such, that it would be absurd to suppose that it can be destroyed, or even lessened, without producing the equivalent of attraction through a given distance of which we have been speaking. You will therefore be surprised to hear that until very recently the universal opinion has been that living force could be absolutely and irrevocably destroyed at any one's option. Thus, when a weight falls to the ground, it has been generally supposed that its living force is absolutely annihilated, and that the labour which may have been expended in raising it to the elevation from which it fell has been entirely thrown away and wasted, without the production of any permanent effect whatever. We might reason, *à priori*, that such absolute destruction of living force cannot possibly take place, because it is manifestly absurd to suppose that the powers with which God has endowed matter can be destroyed any more than that they can be created by man's agency; but we are not left with this argument alone, decisive as it must be to every unprejudiced mind. The common experience of every one teaches him that living force is not *destroyed* by the friction or collision of bodies. We have reason to believe that the manifestations of living force on our globe are, at the present time, as extensive as those which have existed at any time since its creation, or, at any rate, since the deluge —that the winds blow as strongly, and the torrents flow with equal impetuosity now, as at the remote period of 4000 or even 6000 years ago; and yet we are certain that, through that vast interval of time, the motions of the air and of the water have been incessantly obstructed and hindered by friction. We may conclude, then, with certainty, that these motions of air and water, constituting living force, are not *annihilated* by friction. We lose sight of them, indeed, for a time; but we find them again reproduced. Were it not so, it is perfectly obvious that long ere this all nature would have come to a dead standstill. What, then, may we inquire, is the cause of this apparent anomaly? How comes it to pass that, though in almost all natural phenomena we witness the arrest of motion and the apparent destruction of living force, we find that no waste or loss of living force has actually occurred? Experiment has enabled us to answer these questions in a satisfactory manner; for it has shown that, wherever living force is *apparently* destroyed, an equivalent is produced which in process of time may be reconverted into living force. This equivalent is *heat*. Experiment has shown that wherever living force is apparently destroyed or absorbed, heat is produced. The most frequent way in which living force is thus converted into heat is by means of friction. Wood rubbed against wood or against any hard body, metal rubbed against metal or against any other body—in short, all bodies, solid or even liquid, rubbed against each other are invariably heated, sometimes even so far as to become red-hot. In all these instances the quantity of heat produced is invariably in proportion to the exertion employed in rubbing the bodies together—that is to the living force absorbed. By fifteen or twenty smart and quick strokes of a hammer on the end of an iron rod of about a quarter of an inch in diameter placed upon an anvil an expert blacksmith will render that end of the iron visibly red-hot. Here heat is produced by the absorption of the living force of the descending hammer in the soft iron; which is proved to be the case from the fact that the iron cannot be heated if it be rendered hard and elastic, so as to transfer the living force of the hammer to the anvil.

"The general rule, then, is, that wherever living force is *apparently* destroyed, whether by percussion, friction, or any similar means, an exact equivalent of heat is restored. The converse of this proposition is also true, namely, that heat cannot be lessened or absorbed without the production of living force, or its equivalent attraction through space. Thus, for instance, in the steam-engine it will be found that the power gained is at the expense of the heat of the fire,— that is, that the heat occasioned by the combustion of the coal would have been greater had a part of it not been absorbed in producing and maintaining the living force of the machinery. It is right, however, to observe that this has not as yet been demonstrated by experiment. But there is no room to doubt that experiment would prove the correctness of what I have said; for I have myself proved that a conversion of heat into living force takes place in the expansion of air, which is analogous to the expansion of steam in the cylinder of the steam-engine. But the most convincing proof of the conversion of heat into living force has been derived from my experiments with the electromagnetic engine, a machine composed of magnets and bars of iron set in motion by an electrical battery. I have proved by actual experiment that, in exact proportion to the force with which this machine works, heat is abstracted from the electrical battery. You see, therefore, that living force may be converted into heat, and that heat may be converted into living force, or its equivalent attraction through space. All three, therefore—namely, heat, living force, and attraction through space (to which I might also add *light*, were it consistent with the scope of the present lecture)—are mutually convertible into one another. In these conversions nothing is ever lost. The same quantity of heat will always be converted into the same quantity of living force. We can therefore express the equivalency in definite language applicable at all times and under all circumstances. Thus the attraction of 817 lb. through the space of one foot is equivalent to, and convertible into, the living force possessed by a body of the same weight of 817 lb. when moving with the velocity of eight feet per second, and this living force is again convertible into the quantity of heat which can increase the temperature of one pound of water by one degree Fahrenheit. The knowledge of the equivalency of heat to mechanical power is of great value in solving a great number of interesting and important questions. In the case of the steam-engine, by ascertaining the quantity of heat produced by the combustion of coal, we can find out how much of it is converted into mechanical power, and thus come to a conclusion how far the steam-engine is susceptible of further improvements. Calculations made upon this principle have shown that at least ten times as much power might be produced as is now obtained by the combustion of coal. Another interesting conclusion is, that the animal frame, though destined to fulfil so many other ends, is as a machine more perfect than the best contrived steam engine—that is, is capable of more work with the same expenditure of fuel.

"Behold, then, the wonderful arrangements of creation. The earth in its rapid motion round the sun possesses a degree of living force so vast that, if turned into the equivalent of heat, its temperature would be rendered at least 1000 times greater than that of red-hot iron, and the globe on which we tread would in all probability be

MATTER, LIVING FORCE AND HEAT.

rendered equal in brightness to the sun itself. And it cannot be doubted that if the course of the earth were changed so that it might fall into the sun, that body, so far from being cooled down by the contact of a comparatively cold body, would actually blaze more brightly than before in consequence of the living force with which the earth struck the sun being converted into its equivalent of heat. Here we see that our existence depends upon the *maintenance* of the living force of the earth. On the other hand, our safety equally depends in some instances upon the *conversion* of living force into heat. You have, no doubt, frequently observed what are called *shooting-stars*, as they appear to emerge from the dark sky of night, pursue a short and rapid course, burst, and are dissipated in shining fragments. From the velocity with which these bodies travel, there can be little doubt that they are small planets which, in the course of their revolution round the sun, are attracted and drawn to the earth. Reflect for a moment on the consequences which would ensue, if a hard meteoric stone were to strike the room in which we are assembled with a velocity sixty times as great as that of a cannon-ball. The dire effects of such a collision are effectually prevented by the atmosphere surrounding our globe, by which the velocity of the meteoric stone is checked and its living force converted into heat, which at last becomes so intense as to melt the body and dissipate it into fragments too small probably to be noticed in their fall to the ground. Hence it is that, although multitudes of shooting-stars appear every night, few meteoric stones have been found, those few corroborating the truth of our hypothesis by the marks of intense heat which they bear on their surfaces.

"Descending from the planetary space and firmament to the surface of our earth, we find a vast variety of phenomena connected with the conversion of living force and heat into one another, which speak in language which cannot be misunderstood of the wisdom and beneficence of the Great Architect of nature. The motion of air which we call *wind* arises chiefly from the intense heat of the torrid zone compared with the temperature of the temperate and frigid zones. Here we have an instance of heat being converted into the living force of currents of air. These currents of air, in their progress across the sea, lift up its waves and propel the ships; whilst in passing across the land they shake the trees and disturb every blade of grass. The waves by their violent motion, the ships by their passage through a resisting medium, and the trees by the rubbing of their branches together and the friction of their leaves against themselves and the air, each and all of them generate heat equivalent to the diminution of the living force of the air which they occasion. The heat thus restored may again contribute to raise fresh currents of air; and thus the phenomena may be repeated in endless succession and variety.

"When we consider our own animal frames, 'fearfully and wonderfully made,' we observe in the motion of our limbs a continual conversion of heat into living force, which may be either converted back again into heat or employed in producing an attraction through space, as when a man ascends a mountain. Indeed the phenomena of nature, whether mechanical, chemical, or vital, consist almost entirely in a continual conversion of attraction through space, living force, and heat into one another. Thus it is that order is maintained in the universe—nothing is

MATTER, LIVING FORCE AND HEAT.

deranged, nothing ever lost, but the entire machinery, complicated as it is, works smoothly and harmoniously. And though, as in the awful vision of Ezekiel, 'wheel may be in the middle of wheel,' and every thing may appear complicated and involved in the apparent confusion and intricacy of an almost endless variety of causes, effects, conversions, and arrangements, yet is the most perfect regularity preserved—the whole being governed by the sovereign will of God.

"A few words may be said, in conclusion, with respect to the real nature of heat. The most prevalent opinion, until of late, has been that it is a *substance* possessing, like all other matter, impenetrability and extension. We have, however, shown that heat can be converted into living force and into attraction through space. It is perfectly clear, therefore, that unless matter can be converted into attraction through space, which is too absurd an idea to be entertained for a moment, the hypothesis of heat being a substance must fall to the ground. Heat must therefore consist of either living force or of attraction through space. In the former case we can conceive the constituent particles of heated bodies to be, either in whole or in part, in a state of motion. In the latter we may suppose the particles to be removed by the process of heating, so as to exert attraction through greater space. I am inclined to believe that both of these hypotheses will be found to hold good,— that in some instances, particularly in the case of *sensible* heat, or such as is indicated by the thermometer, heat will be found to consist in the living force of the particles of the bodies in which it is induced; whilst in others, particularly in the case of *latent* heat, the phenomena are produced by the separation of particle from particle, so as to cause them to attract one another through a greater space. We may conceive, then, that the communication of heat to a body consists, in fact, in the communication of impetus, or living force, to its particles. It will perhaps appear to some of you something strange that a body apparently quiescent should in reality be the seat of motions of great rapidity; but you will observe that the bodies themselves, considered as wholes, are not supposed to be in motion. The constituent particles, or atoms of the bodies, are supposed to be in motion, without producing a gross motion of the whole mass. These particles, or atoms, being far too small to be seen even by the help of the most powerful microscopes, it is no wonder that we cannot observe their motion. There is therefore reason to suppose that the particles of all bodies, their constituent atoms, are in a state of motion almost too rapid for us to conceive, for the phenomena cannot be otherwise explained. The velocity of the atoms of water, for instance, is at least equal to a mile per second of time. If, as there is reason to think, some particles are at rest while others are in motion, the velocity of the latter will be proportionally greater. An increase of the velocity of revolution of the particles will constitute an increase of temperature, which may be distributed among the neighbouring bodies by what is called *conduction*—that is, on the present hypothesis, by the communication of the increased motion from the particles of one body to those of another. The velocity of the particles being further increased, they will tend to fly from each other in consequence of the centrifugal force overcoming the attraction subsisting between them. This removal of the particles from each other will constitute a new condition of the body—it will enter into the state of fusion, or become melted. But, from what we have already stated, you will perceive that in order to remove the particles violently attracting one another asunder, the expenditure of a certain amount of living force or heat will be required. Hence it is that heat is always absorbed when the state of a body is changed from solid to liquid, or from liquid to gas. Take, for example, a block of ice cooled down to zero; apply heat to it, and it will gradually arrive at 32°, which is the number conventionally employed to represent the temperature at which ice begins to melt. If, when the ice has arrived at this temperature, you continue to apply heat to it, it will become melted; but its temperature will not increase beyond 32° until the whole has been converted into water. The explanation of these facts is clear on our hypothesis. Until the ice has arrived at the temperature of 32° the application of heat increases the velocity of rotation of its constituent particles; but the instant it arrives at that point, the velocity produces such an increase of the centrifugal force of the particles that they are compelled to separate from each other. It is in effecting this separation of particles strongly attracting one another that the heat applied is *then* spent; not in increasing the velocity of the particles. As soon, however, as the separation has been effected, and the fluid water produced, a further application of heat will cause a further increase of the velocity of the particles, constituting an increase of temperature, on which the thermometer will immediately rise above 32°. When the water has been raised to the temperature of 212°, or the boiling-point, a similiar phenomenon will be repeated; for it will be found impossible to increase the temperature beyond that point, because the heat then applied is employed in separating the particles of water so as to form steam, and not in increasing their velocity and living force. When, again, by the application of cold we condense the steam into water, and by a further abstraction of heat we bring the water to the solid condition of ice, we witness the repetition of similar phenomena in the reverse order. The particles of steam, in assuming the condition of water, fall together through a certain space. The living force thus produced becomes converted into heat, which must be removed before any more steam can be converted into water. Hence it is always necessary to abstract a great quantity of heat in order to convert steam into water, although the temperature will all the while remain exactly at 212°; but the instant that all the steam has been condensed, the further abstraction of heat will cause a diminution of temperature, since it can only be employed in diminishing the velocity of revolution of the atoms of water. What has been said with regard to the condensation of steam will apply equally well to the congelation of water.

"I might proceed to apply the theory to the phenomena of combustion, the heat of which consists in the living force occasioned by the powerful attraction through space of the combustible for the oxygen, and to a variety of other thermo-chemical phenomena; but you will doubtless be able to pursue the subject further at your leisure.

"I do assure you that the principles which I have very imperfectly advocated this evening may be applied very extensively in elucidating many of the abstruse as well as the simple points of science, and that patient inquiry on these grounds can hardly fail to be amply rewarded."

In order to duly appraise work, it is necessary to realise the obstacles overcome, and this may be very difficult. To those who work in the light it is almost impossible to conceive the difficulties encountered by those working in the dark. So, looking back, after the general revolution in philosophical thought caused by the recognition of the conservation of energy, it is very hard to arrive at a just estimate of the work of those by whom it was effected.

To the present student, endeavouring to enter fully into the history of the discovery, difficulties are presented by his own familiarity with expressions and terms, now part of his language, but which came into existence during the discovery, by his familiarity with those facts the discovery of which resulted in the generalization, and by his familiarity with facts almost innumerable, previously unknown, but which have been revealed as a consequence of the generalization.

The terms 'energy' and 'work' did not exist in the language of science in their present significance. The *vis viva* of a body, the product of the square of its velocity multiplied by its mass, had since the time of Newton been recognized as a mechanical quantity, and the term 'energy' had been applied to the half of this quantity by Young. On the other hand, 'work'—motion against resistance—expressed as the product of the distance, multiplied by the mean resistance overcome, although it was known to express the half of the change in the *vis viva* which takes place in a body moving against resistance, had never been recognized in the schools of mechanical philosophy as a fundamental measure of mechanical action, either as 'work' or by any other name.

Outside the schools of mechanical philosophy, engineers engaged in constructing and using the steam engine had long been led to recognize motion against resistance as the mechanical and commercial measure of potency, and under the names of 'work' and 'accumulated work,' these men had become familiar with what are now known as 'work' and 'energy' (actual and potential). But in so far as 'work' was spent in overcoming friction, it was rightly considered as the measure of mechanical power annihilated.

Heat had been recognized as a measurable quantity, expressed by product of the weight of water multiplied by the rise of temperature effected by the heat. The discoveries of Black and Watt had led to the recognition of something in heat other than that which merely went to raise the temperature of matter, a recognition expressed in the term caloric, invented by Lavoisier (1787), for the imponderable fluid which imparted elasticity and temperature to matter, and which in Lavoisier's view might be *vis viva*. But suggestive, as this recognition undoubtedly was, to those physical philosophers who directed their philosophy to the subject, the orthodox school still maintained, as a dogma, the materiality of heat, and ignoring Lavoisier's reservation as to the materiality of caloric, continued to regard heat as unchangeable, like a definite form of matter, accepting all the physical casuistry which had been invented in order to reconcile this dogma with observed phenomena. That the development of heat by friction between bodies in relative motion was evidence that sensible heat consisted of some form of mechanical activity, set up by the action between the bodies, had been borne in upon the minds of philosophers from very early times, and particularly during the previous 200 years. This view, however, had come to be regarded as opposed to the dogma of caloric by those who held it as well as by the orthodox school (*See* Appendix); the idea of the materiality of heat associated with the dogma

MATERIALITY OF HEAT.

having become so general, that it does not appear to have occurred to any philosopher, before this time, that the mechanical effect which the mechanical action of overcoming friction (annihilating mechanical power) imparted to the bodies, could in itself be the very imponderable fluid, measurable, indestructible, and uncreatable, that constituted caloric. That such a simple reconciliation, as it now appears, should have been overlooked by both schools, receives its explanation in the fact, already mentioned, that in the schools of mechanical philosophy, "Work," the only general measure of the mechanical effect transmitted during mechanical action, had not been recognized; so that those who realized that heat was the result of mechanical action were yet unaware that the action represented a transmission to the heated body of a measurable amount of mechanical effect, and thus failed to look to the measure of this effect as affording a measure of the heat produced.

Rumford came very near to revealing the true nature of the relation between heat and mechanical effect by observing that two horses working steadily against a frictional resistance produced heat at a steady rate; but although he measured the quantity of heat produced, and even went so far as to compare this with the heat that would have resulted from the combustion of the food consumed by the horses, he did not attempt any discussion of the mechanical action involved. So that until 'work' was recognized in the schools as the measure of mechanical potency, the description of his investigation did not apparently contain a logical account of the relation between heat and mechanical action, however fully Rumford himself may have realized such a relation.

The condensing steam engine had been at work for 150

years converting heat into mechanical effect, but, so far from suggesting such a conversion, it seemed to indicate the contrary. For not only did all experience show that the power which was developed by the agency of heat was ultimately annihilated in overcoming friction, but the condenser, by which the heat was removed from the engine, was as essential to its action as the furnace in which the heat was produced. And although, as we now know, the heat removed by the condenser is less than the heat received from the boiler by the equivalent of the work done, this latter quantity is so small compared with either of the former that it can only be detected by most difficult tests, which not only had not been made at that time but were not made for 20 years after its existence was known as the result of the discovery of the law of the conservation of energy.

Nor was this all the apparently contrary evidence afforded by the steam engine. It rendered it evident that a principal function of the heat received was that of imparting elasticity to the steam, which elasticity having been expended in doing work, apparently left the steam possessed of its heat but useless. The amount of work that could be obtained from a given amount of heat was thus found to depend on the degree of elasticity or pressure at which the water was boiled, and this depended on the temperature; so that the steam engine, so far from showing that heat was converted into work, showed that the amount of work produced by the agency of heat depended on the temperature. This experience was generalized by Carnot, 1824, in his well-known Theorem, which, though perfectly true and consistent with the conversion of heat, was for some time an apparent stumbling block.

THE LOCOMOTIVE OBTRUSIVE.

The condensing engine, however, contributed to the discovery of the mechanical origin of heat, in that it led to the recognition of work as the measure of mechanical action ; and to the locomotive must be attributed the birth of that philosophical interest respecting heat and work which immediately followed its general introduction. The condensing engine had not been obtrusive—it was not generally to be seen unless looked for. The locomotive is obtrusive ; it will be seen : and by 1842 locomotives had obtruded themselves pretty well all over Europe. They immediately took their places as objects of as much wonder and interest to the grown people who saw them for the first time as they are still to the young ; demanding the attention even of philosophers who had previously studied nothing lower than the planets. It thus turned out, as became known some eight years after, that about this time three philosophers, Séguin (1839), Mayer (1842), and Colding (1843), suggested, as the result of philosophical reasoning, that heat is convertible into mechanical effect.

But it was not by such suggestions that the prejudice and casuistry respecting caloric were to be removed. Even had these suggestions met with general acceptance, had the convertibility of heat into mechanical effect been recognized at the time, it does not follow that the result would have been the general recognition of the conservation of energy. Besides mechanical effect, there were the other physical effects which result in combustion and the passage of electricity in conductors, by which heat is produced or developed, as it was then called, and these were then even less understood. And although Mayer followed up his first suggestion by developing many of its more important consequences, his work attracted little notice and carried no conviction.

CHEMICAL AND PHYSICAL EFFECTS.

The interest, philosophical and practical, taken in matters electrical at that time was intense, and far outweighed that taken in heat. The discovery of Oersted, in 1820, that the compass needle is deflected from its usual direction by an electric current parallel to the needle, followed by Sturgeon's discovery, in 1825, of the soft iron electro-magnet, had, in 1837, rendered the electric telegraph practical in the hands of Cooke and Wheatstone. Ohm had, in 1827, discovered the relation between the electromotive force of the battery, the resistance of the circuit, and the current generated. Faraday had discovered the current caused in a closed circuit by the motion of a magnet—"magneto electric induction"—and was engaged in his now classical "Experimental Researches in Electricity," having already shown that the quantity of electricity produced in the battery is proportional to the number of chemical equivalents electrolyzed. In 1836, Sturgeon added a new interest to the subject by his invention of the "Commutator," and his construction by its means of two machines, one the magneto-electric machine, now called the "dynamo," and the other the electro-magnetic engine, now called the "motor." By the former of these, a current of electricity, having apparently all the properties of a voltaic current, was produced by expending mechanical effect in turning an electro-magnet between the poles of a stationary magnet; while by the second, a voltaic current from a battery, acting on the coil of an electro-magnet turning between the poles of a stationary magnet, urges it to overcome resistance and produce mechanical effect. These machines played an important part in the discovery of the law of conservation of energy. By the first, an electric current is produced solely by the expenditure of mechanical effect, evincing the convertibility of the mechanical effect expended into the effect of the current which it produced. By the second, mechanical effect is produced as the effect of the electric current, evincing with the first machine the reciprocal convertibility of the two effects; and besides this, since the current originates in the zinc electrolyzed in the battery, showing the convertibility of the effect of chemical combination as effected by electrolysis and the mechanical effect produced.

Clear and direct as this evidence now appears, it was by no means clear in 1838, as is shown by the fact, that, except by one "man," it was not definitely recognized for twelve years, and then only as the result of this man's teaching, no one in the meantime having perceived it.

This obscurity was, doubtless, owing to the general interest in electricity at that time being in the phenomenal effects rather than the quantitative, and in the absence of any common language in which to express the latter. Quantitative expressions for electricity had been used by various investigators, but they were not by any means uniform, each investigator using his own expressions, which were little known; the significance of the quantitative measures being still less apprehended. Thus Ohm had shown that the electrical action of a battery in overcoming the electrical resistance of the circuit, was always proportional to the product of the quantity of current multiplied by the electromotive force of the elements used. Faraday had shown that the quantity of current was proportional to the number of chemical equivalents electrolyzed. Thus, from Ohm and Faraday together, it was to be inferred that the electric action of a battery in overcoming the resistance of the circuit is proportional to the product of the electromotive force of the elements multiplied by the number of elements electrolyzed. The inference, however, was not drawn. Again, it was well known that heat was generated by a current in a conductor, and it was generally believed that the heat was proportional to the electric action expended, expressed by Ohm as the product of the electromotive force multiplied by the current, but no definite experiments had been made to establish this; nor had there been any suggestion of the inference which immediately followed, that the heat developed in the entire circuit of the battery would be proportional to the same product which expressed the electric action of the battery in overcoming resistance, the number of equivalents of the elements combined multiplied by the electromotive force of these elements. Still less had it been suggested that this definite development of heat, associated with the chemical combination effected by the electrolysis of definite quantities of definite elements, was identical with the quantity of heat that would be produced by the chemical combination of the same elements—the heat generated by combustion.

The practical possibilities apparently underlying Sturgeon's invention as a prime-mover which might displace the steam engine, secured attention to the exclusion of the philosophical significance of the actions revealed. How to increase the power and economy of the engines? was the question for the time. The heat generated by the currents in overcoming the resistance of the conductors, and that generated by the friction of the working parts, being looked upon as necessary evils to be minimized, rather than as keys by which the most general law in the universe was shortly to be revealed.

CHAPTER II.

PARENTAGE AND EARLY LIFE.—*Sees the First Train.— Education and Companionship with his Brother.— Association with Dalton.—Similarity of the Works of Dalton and Joule.—The Brothers' Vacations.—Their Amusements.—Under Treatment for the Spine.—Commencement of Life.—Continued Companionship.—Visits to Dalton.—Further Instruction in Chemistry.—Intercourse with Sturgeon and Members of this Society.— Joule a Dangerous Companion with a Gun.—British Association.—Dr. Scoresby.—Visit to Bradford.—Interruption of Companionship.—Joule's Activity as a Boy.*

James Prescott Joule was born on the 24th of December, 1818, at Salford. His grandfather, William Joule, or, as, sometimes written in old deeds, Youle, was born in June, 1745, at Youlgreave, in Derbyshire, where the family were yeomen. He migrated to Salford, and established a brewery concern. He acquired wealth, and was much respected. His death took place at Buxton, 21st May, 1799; an obituary notice appearing in Harrop's *Manchester Mercury*. Joule's father, Benjamin Joule, fourth son of William, born in Salford, 9th March, 1784, married on the second of May, 1814, Alice, daughter of Thomas Prescott, of Wigan; she died 20th December, 1834, aged 48. They had five children, of whom Joule was second, the present Mr. Benjamin St. J. B. Joule being the eldest, the others were John, and two daughters—Mary and Alice. Joule's father, who was an invalid for nine years before his death, died at the age of seventy-four, in 1858. He was wealthy, owning the Salford Brewery, which he sold in 1854. Joule himself never took any active part in the management of the brewery.

Joule and his elder brother were educated at their father's house, Broom Hill, near Manchester. They were constant companions and greatly attached to each other. Joule was delicate and under treatment for the spine. This attachment lasted till Joule's death (1889), as did also their companionship with interruptions. Mr. B. St. J. B. Joule kept a diary from early life, and has kindly furnished the author with all the extracts referring to his brother.

The first entry refers to their going, on September 15, 1830 (a day memorable for the opening of the Manchester and Liverpool Railway and the death of Mr. Huskisson, who expired in the Parsonage at Eccles), "into a field near Eccles to see the first trains which travelled between Liverpool and Manchester, and to their riding on several Saturday afternoons to a place between Eccles and Patricroft to watch the two trains (one on each set of rails) passing and repassing for the amusement of passengers to Newton-in-the-Willows and back." Joule was then eleven years of age; and the brothers were under the private tuition of a resident master, S. T. Porter. Their next tutor was Mr. Frederick Tappenden, who came from a military school in the south, December 17, 1832, and left them in December, 1834. They rode their ponies together and played tricks together; among others passing electric shocks on friends and servants, making them as strong as possible with the poor apparatus then in vogue, and encouraging their subjects by standing in line with them and professedly receiving the same shocks, though in reality passing the electricity through a hidden bye wire. They also brought down electricity by kites, and though they took precautions, found it a game not to be trifled with.

In 1834 their father determined that they should study chemistry under Dalton, who received pupils in the buildings of this Society, 36, George Street, of which he was at the time still president. Dalton required that his pupils should be well grounded in arithmetic and the first book of Euclid before admitting them. Mr. Tappenden accordingly prepared them before he left in 1834, and introduced them to Dalton, who, however, took no notice of him. To their dismay, not to say disgust, at the end of an hour, they had not got through addition. They attended, with some short breaks, twice a week for one hour, and at the end of two years had just got through arithmetic and the first book of Euclid when Dalton suggested that they should proceed to the higher mathematics. This they declined to do, and were only commencing chemistry when Dalton had a severe attack of paralysis, April 18th, 1837, which cut short their periodical visits for instruction.

Thus commenced and ended Joule's first association with this Society. His early connection with the great master who had founded the science of Modern Chemistry must be considered one of the happiest circumstances of Joule's life. Although advancing years and weakening powers had diminished the fulness and vigour of Dalton's instruction, still the familiar intercourse with a man so simple yet so profound, successful, and distinguished, must have strengthened Joule's natural bent to observe for himself, and to trust implicitly the conclusions to which he was led by his own observation. Besides which, the opportunity afforded by Dalton's laboratory of becoming intimate with the simple but effective home-made apparatus by which Dalton had accomplished his great work, could not but encourage young Joule in that path of self-dependence in his experimental work, which is a most remarkable feature of his career, and which contributed so greatly to his success. There can be no doubt, however, that Joule possessed constructive genius and powers of manipulation far beyond those of his master, whose genius lay rather in the observation of nature and philosophical reasoning.

It would be perhaps going too far to infer that Joule unconsciously took the leading idea which guided him to his discoveries from Dalton or his work, but it is nevertheless remarkable that the chief distinction between the experimental work of these men, each in his own line, and that of their early contemporaries, was the same, namely, the substitution of quantitative measurement for mere phenomenal experiments. Dalton determined the chemical equivalents by weighing and comparing the weights to the equivalent weight of hydrogen, thus discovering the Atomic Theory and introducing quantitative analysis into chemistry. While Joule, by measuring the several physical and mechanical effects by definite standards, and comparing each with the equivalent electrical effect, arrived at their general equivalence and the law of the universal conservation of energy.

The young Joules continued to visit Dalton after his illness. He was pleased to see them as well as by the occasional information which they gave him, as resulting from their own observation; *e.g.*, from their father's house at Broom Hill they could see as far as Runcorn, and on one fine evening they fancied they saw flashes of lightning in that direction close to the horizon, and that they heard thunder, and observing carefully they found that the sounds followed the flashes at the same intervals of time. On reporting this to Dalton he told them that the greatest distance he had ever heard sounds was some artillery at Liverpool which he heard near Bury ; and subsequently he informed them that he had ascertained that at the time they named a great thunderstorm had been encountered at sea, 40 miles south of Holyhead.

In September, 1835, the brothers were in the Lake District for a few days, and commenced a tour, ascending Helvellyn and Skiddaw. 18th November, they observe—" A most splendid aurora—a strong wind—heavy clouds—no moon—yet we could read ordinary print. At 8 o'clock p.m., for about a minute there were perpendicular beams between the clouds, with primitive colours; at 9 o'clock, the scene was grand ; beams constantly flashing a little S.E. from every quarter of the heavens, and shortly after, there was a bright arc from magnetic east to magnetic west."

On 18th March, 1836, the brothers were at Todmorden—Joule under treatment for the spine by the brothers Taylor ; nevertheless they daily ascended the highest hills in the neighbourhood though they were covered with snow. On 15th May, they observe and record an account of the eclipse of the sun.

In 1838, the diary contains the following entries :—" April 12th.—We called on Dr. Dalton, who seemed very pleased to see us." " April 16th.—A snow storm, James and I did not ride our ponies to-day." " May 8th—11th.—With James at Dr. Taylor's, Todmorden." "June 8th.—James and I rowed to Lowood Inn (Windermere) and back before breakfast, ascended Helvellyn from Wyburn; much snow; played at snow-balling on the summit." "June 11th.—(Keswick) James and I rowed to the end of Derwent-Water and back before breakfast." "June 12th.— A rowing race; we beat two of the best rowers in the district; in the afternoon ascended Skiddaw."

Joule had now a room in his father's house at Broom Hill, for a laboratory, where he had already, though all unconscious, commenced the research which was to end so successfully, while the elder brother was occupied as his father's political agent. The brothers, however, still continued their companionship, riding their ponies, calling on Dalton, who always seemed pleased to see them and invited them to tea. In 1839 they commenced taking private lessons in chemistry from John Davies, and later in the year attended Mr. Davies' lectures given to the medical students; and in the summer of that year they are amusing themselves at York; rowing on the Ouse. In 1840 James made experiments on a lame cart-horse with galvanism. They escort Dalton round their father's brewery; June 11th and 13th were at Matlock New Bath; and at Malvern October 15th ascending the highest hill before breakfast. In 1841 James was very busy with his experiments; they accompanied Peter Clare, secretary of the Literary and Philosophical Society, to see Sturgeon fly an electric kite; they also attended some of the meetings of the Society.

At this time it appears that several of the leading members of the Society were on terms of social intimacy at Broom Hill, as there is an entry May 3rd, 1842, "John Davies, J. H. Ransome, and W. Sturgeon to dinner at father's." On the 19th the family were at Windermere, where the brothers sound all over the lake and find the greatest depth to be 33 fathoms—ascend all the hills, including Scafell, making observations. They have guns and pistols, of which they are very fond, having had one made for them in Birmingham with a specially small bore. James is rash with them, as the following extract shows:—" May 24, Lake Windermere.—After breakfast our party were rowed by James and myself to one of the islands. I then rowed James a short distance away to let them hear a very good echo which we had discovered. I was not observing what James was doing, though I thought he was unusually long in loading the pistol (a large old-fashioned cavalry weapon, used by my father when he belonged to the Manchester Mounted Volunteers), when I was suddenly startled by a tremendous report, and on looking round found that the pistol had disappeared. The "knock" had been so violent that the weapon had been wrested out of James' hand and had fallen into the lake. He told me that he wished to produce the loudest report possible, and had used three charges of powder."

It seems that James was anything but a desirable companion with a gun, being very absent and playing with his trigger, on one occasion shooting off his own eyebrows, and affording his brother many narrow escapes. James finds amusement in painting, having considerable skill. He is also fond of pictures, giving, at this time, as much as £50 for a cattle piece. The elder brother has great taste for and proficiency in music. James also is fond of music, and has some small proficiency on the piano. The British Association meets in Manchester in 1842, and the Revd. Dr. and Mrs. Scoresby are entertained at Broom Hill, and afterwards the brothers return the visit to the parsonage at Bradford, travelling by rail to Brighthouse, each riding on the roof of a first-class carriage, as they frequently did. They attended morning service at Bradford, and in the afternoon accompanied Dr. and Mrs. Scoresby to service at Shipley, where they sat with Mrs. Scoresby in the parson's pew, but James did not forget his rule of sleeping during the afternoon sermon, though his host was the preacher.

In the autumn they are spending what is to be for some time their last vacation together in Wales, respecting which there are the following entries :—" September 17th.—Drive nine miles of the road from Carnarvon to Bedgellert, and as James agreed with me that there was a practicable way we left the carriage and arrived at the top in less than two hours. 21.—Again ascended Snowdon without guide from Bedgellert and ascended by what we thought was a short route; it occupied the same time."

On the 7th of December, 1842, Mr. B. St. J. B. Joule was married, which for a short period disturbed the close alliance of the brothers.

From the foregoing account it appears that during his vacations and in ordinary matters young Joule, until the age of 23, was possessed by the interests and enjoyed the pursuits of a healthy, vigorous boy, just entering on his manhood. This is important, since, judging from his work during the last four years of this period, it appears that he possessed in no ordinary degree the attributes of sober and experienced manhood, having by the end of 1842 acquired, in almost every branch of physical science, knowledge in many respects far beyond that of any philospher then living; and having, at the age of 24, discovered all but one of the relations between the various physical effects which disclosed the law of the conservation of energy.

CHAPTER III.

JOULE'S FIRST RESEARCH.—*Starts to Improve Sturgeon's Electro-Magnetic Engine. — Increases the Magnetic Force. — Does not Realize His Problem. — Effects the Absolute Measure of 'Work.'—Finds that the Speed is Limited.—Seeks for the Limit in the Magnets.—Fails to Find It. — Realizes the Importance of Measuring the Current. — Constructs a Standard Galvanometer.— Repeats His Experiments.—Discovers Fundamental Law of Electro-Magnetic Attraction.—Contemplates Perpetual Motion.—Explains Law of Electro-Magnetic Attraction.— Measures Current, Velocity, Resistance, and Estimates Zinc Consumed in Producing the Current. — Obtains 'Duty' per lb. of Zinc. — Realizes Resistance to the Current Induced by the Motion of the Magnet.—Refers to Faraday, Ohm, &c. — Determines Law of Induced Resistance.—Discovers Equivalence of Mechanical Effect to the Electric-Action, and Chemical-Action Expended in its Production.—Introduces Absolute Electric Measurement.—Concludes that the Electro-Magnetic Engine can never Compete with the Steam Engine. — Sees a great Philosophical Discovery before him.*

It was with the simple object of improving Sturgeon's electro-magnetic engine that Joule, in 1838, then 19 years of age, entered upon the first stage of his great research, publishing the results in letters to Sturgeon's "Annals of Electricity." He was impressed, like many older men, with the apparent possibility "that electro-magnetism will ultimately be substituted for steam in propelling machinery." That he had any underlying philosophical motive in no way appears. That he already possessed considerable familiarity with the voltaic and electrical appliances then in vogue is clear, but this is about all. He does not seem to have, at this time, studied the philosophy of the subject, or to have had any special acquaintance with the work of Oersted, Ohm, or Faraday. Of mechanical philosophy he appears to have had no knowledge beyond that which was familiar to the practical engineers amongst whom he lived, and what had resulted from his own observation. That he possessed the engineer's rather than the philosopher's knowledge of mechanics proved one of the happiest circumstances; as he was thus familiar with the only measure of mechanical action which directly measured the mechanical effect of the physical actions he was about to study.

As the purpose of the electro-magnetic engine, like that of all prime movers, is to cause motion against resistance, Joule was at once introduced to the measurement of the physical actions in his engine in terms of 'work,' the only measure of mechanical effect, but he did not at first apprehend this in its full significance. The first thing to catch his attention was how to increase the force which the engine would exert. The "dynamo" is now, as the "electro-magnetic engine" was then, essentially feeble as regards effort. This effort depends on the attractive force of the magnets, and Joule's first paper shows that his attention was solely directed to the statical effect—how with a given battery to obtain the greatest amount of force out of the least weight of soft iron and wire? He had constructed a compound magnet, consisting of small soft iron horse shoe magnets of definite proportions, with their poles in the same plane in close order, which he finds "gives good lifting power," the reasons for which he states. To describe his success so far, and explain how he proposes to adapt this to the machine, constitutes his first paper, 8th January, 1838. The problem before him is, however, not how to get the greatest lifting force out of his magnets, but how to get the greatest amount of mechanical work with the least weight of material out of a given battery. He has not, apparently, paid any attention to the problem of the battery. His second paper, December 1st, 1838, shows that when he sets his engine to work he is face to face with the other factor of work—motion. His engine has force but not speed. "I finished the engine I was working upon last summer." It had thus occupied him six months, a period which will not seem long, considering that he did all the work himself, and did it with the utmost care. In the next sentence he says: "It weighs 7½ lbs., and the greatest power I have been able to develop with a battery of forty-eight Wollaston four-inch plates was to raise 15 lbs. a foot high in a minute, in which the friction of the working parts, which was very considerable, was reckoned." In this sentence he uses '*power*' in its accepted sense, rate of motion against resistance; and though he does not introduce the expression 'work,' he has effected its *measurement* and used it as expressing the mechanical potency derived from his machine. The passage is thus remarkable, as showing Joule's early appreciation of this fundamental measure of mechanical action. It is also remarkable, as containing the first recorded absolute measurement of work in connection with the philosophical study of physics.

In his next sentence Joule remarks: "The result shows

that the advantages of the close arrangement of electro-magnets are not such as I anticipated." His six months' work had thus ended in complete disappointment as regards the results he expected, but he is not discouraged. Here is where the character of the man, or rather boy, comes in. Being assured of the soundness of the reasoning, as far as it went, on which his anticipations were based, and also of the means he had taken to put these anticipations to the test, he recognises in the result of his six months' work a proof of further causes of which he was ignorant. He is not disillusionised in his view as to the possibilities of the electro-magnetic engine, but he sees that he has not fully mastered the principles on which its improvement depends. "I was desirous," he proceeds, "before attempting to make another engine, to satisfy myself how far it was possible to increase the velocity of rotation, which was only $3\frac{1}{2}$ feet per second in the above trial. Now of the many things which limit the velocity, the resistance which iron opposes to the instantaneous induction of magnetism is of considerable importance. I think I shall be able to show how this may be obviated in some measure." And he then proceeds in this and his third letter, March 27, 1839, to describe experiments with other engines in which he uses wire magnets and batteries of different intensity, and then experiments on the lifting power of various forms from which he obtained somewhat unexpected results, which were of great interest in the study of electro-magnetism.

He had then been at work 14 months from the publication of his first paper, and had apparently found very little to encourage him in the avowed object of his research. It appears, however, from his next paper, that the results he had obtained had led him to make a new departure which must be regarded as a fundamental step in attaining the ultimate success of his research. So far he had made no electrical measurements; he had compared his different magnets by connecting them with the same battery, the only other measurements being as to the effects on the same magnet of varying the number of cells and the charge. His fourth letter, May the 14th, 1839, commences: "It was important to use in the research a galvanometer, the indications of which could be relied upon." This letter is written less than two months after the previous one, yet in all the account of his previous 14 months' work there is no suggestion of even the desirability of measuring the currents of his batteries. In his first two letters he ends by indicating the line in which he intends to proceed; it would therefore seem probable that he had only realized the importance of this step since the 27th of March, 1839. In his fourth letter he proceeds at once to the description of the galvanometer, which is an instrument of his own device and construction, far in advance of anything of the kind existing at that time, and for its purpose as perfect as it would now be possible to make it. That he was fully acquainted with the principles of its action, and had thought deeply as to its construction, is abundantly evident. In this letter, after describing the instrument, he gives no indication as to what is his intention in using it, but proceeds to describe experiments with the magnets previously constructed in which the current is measured with the galvanometer.

The result of these experiments is to reveal an important law up to that time unknown, "that the *attractive force of two electro-magnets for one another is as the square of the product of the current and the length of the wire.*"

The discovery of this law releases him from further investigation on the magnetic attraction in his engines. He recognises this, and the conclusion of his paper is remarkable, as showing how far he is as yet from apprehending the remotest relation between the work which his machine may perform, and the source of work, the consumption of zinc in the battery, and also in showing that he sees no fundamental objection to perpetual motion. "I can hardly doubt," he says, "that electro-magnetism will ultimately be substituted for steam to propel machinery. If the power of the engine is in proportion to the attractive force of its magnets, and if this attraction is as the square of the electric force, the economy will be in the direct ratio of the quantity of electricity, and the cost of working the engine may be reduced *ad infinitum*. It is, however, yet to be determined how far the effect of magnetic electricity may disappoint these expectations."

He then, as usual, indicates the line he intends to follow in his new engine. The next two letters relate to a verification of the law already mentioned, and a description of the new machine, of which he says; "In my preliminary trials I have been much pleased with its performance." This is August 30th, 1839.

There is then a long interval, during which he has been experimenting with his engine, and digesting the results of these and previous experiments. His seventh letter, dated March 10th, 1840, on "Electro-magnetic Forces," contains, in the first instance, a highly philosophical explanation of the law of attraction he had discovered. This is followed by a description of the experiments with the new machine, and a philosophical discussion of the results. He has made a great advance all round. In these experiments he measures

INDUCED RESISTANCE.   39

accurately the relative current with the galvanometer, the resistance and velocity of the machine, obtaining what he now calls *work*, and perhaps most important, he now estimates that portion of the zinc consumed in his batteries, which is effective in producing current; thus obtaining the work done per pound of zinc, which he expresses, after the manner of engineers, as 'duty.' He has thus made complete tests of the machine in every particular, varying the resistance, the velocity, and the quantity and intensity of the current.

All this implies that since his last paper he has received much fresh light; whether or not this is solely the result of his own researches, is not stated in the paper, but from the manner in which he commences the discussion of the results, it may be inferred that it is not.

The results, as stated, show clearly, that although the resistance overcome is as the square of the current in all cases, verifying his law of attraction as applied to the moving armature, the current and resistance overcome diminish as the speed increases—indicating clearly the induced resistance (Faraday's) in the wire caused by the motion of the machine, showing him that with a constant battery the force of his engine diminishes as the speed increases; thus revealing the fundamental limit to speed. Considering that he, in consequence of not being aware of this induced resistance, or having over-looked it, has been labouring for more than two years trying to realize impossibilities, and that at the very first start his experiments suggested to him the existence of some limit to speed, he could hardly have discovered this internal resistance for himself without making some comment upon his discovery—whereas what he says is:

"The above examples will show pretty clearly the effects of magnetic internal resistance. This resistance is the prime obstacle to the perfection of the electro-magnetic engine; and in proportion as it is overcome will the motive force increase. It therefore claims our first attention." This suggests that he had become aware that this resistance was previously known. A view, which is confirmed by his attributing, in a subsequent paper, Feb. 16, 1841, the discovery, without determining the laws of this internal resistance, to Professor Jacobi, and also to his giving references to *Faraday's Experimental Researches; Scoresby's Magnetic Researches, Nesbit and Henry*, in a paper, written in August 28, 1840, which are almost the first reference contained in his papers to the works of other investigators, and thus indicate that he had recently been reading as well as investigating. Joule is not, however, content with recognizing this internal resistance, but has made a complete experimental determination of its law—"that it is proportional to the product of the velocity of rotation multiplied by the magnetism." This statement is equivalent to saying that the electric action spent in overcoming the resistance induced in the machine, namely, the product of the current multiplied by the induced resistance, is proportional under all circumstances to the product of the square of the current multiplied by the velocity of the machine; and this is also proportional to the mechanical effect; so that Joule had now proved that there are quantitative equivalents in mechanical effect for the electric action (product of electro-motive force multiplied by current), and for the chemical action expended in producing the mechanical effect.

He then continues his researches on magnetic forces

and in the next paper describes another step in electrical measurement, which, besides being of primary importance in leading him on to his final goal, constitutes him the father of the present system of absolute units in science. This was the adoption of a definite quantitative unit of electricity. The connection between this step and his last experiments may be clearly inferred. In these he wanted to obtain a measure or estimate of the quantity of zinc consumed in supplying the current he was using. The actual consumption in the battery included besides this, that which was wasted by local action through the imperfections of the batteries. To eliminate this wasted zinc, he made use of "Faraday's discovery of the definite quantity of electricity associated with the chemical equivalents of the bodies;" the current caused by the consumption of each equivalent of zinc in the battery being capable of dissociating one equivalent of the hydrogen in the voltameter. To avail himself of this law, his galvanometer "was connected with an apparatus furnished with fine platinum electrodes. A current was transmitted through the instrument, and after a few minutes the current was broken, and the hydrogen measured in a graduated tube." The mean of ten trials showed that "0·76 gr. of water was decomposed per hour by the current indicated by each unit of my former numbers." The decomposition of 9 grains of water, i.e., one grain of hydrogen, would represent the solution of one equivalent (32·3) of zinc, so that the current represented by each degree of his galvanometer represented the solution of 27·27 grains of zinc. He had made these experiments and calculations before March 10th, 1840, as they are included in the *tables* in his 7th letter, but the determination of adopting them as a general quantitative unit of electricity seems to have been taken subsequently. Of this he says, in his 8th letter: "The difficulty, if not impossibility, of understanding experiments, and comparing them with one another, arises, in general, from incomplete description of apparatus, and from the arbitrary and vague numbers which are made to characterise electric currents. Such a practice might be tolerated in the infancy of the science, but in its present state of advancement greater precision and propriety are imperatively demanded. I have, therefore, determined for my own part to abandon my old quantitative numbers, and to express my results in the basis of a unit which shall be at once scientific and convenient. That proposed by Dr. Faraday is, I believe, the only standard of this kind which has been suggested. His discovery of the definite quantity of electricity, associated with the atoms or chemical equivalents of bodies, has induced him to use the voltameter as a measure, and to propose that a hundredth part of a cubic inch of the mixed gases forming water should constitute a *degree*. There can be no doubt that this system would offer great advantages to the experimenter in somes cases, and when the Ohm instrument is employed. However, as I am not aware that it has been used in the researches of any electrician, not excepting those of Faraday himself, I do not hesitate to advance what I think more appropriate as well as more generally advantageous. It is thus simply stated ":—

1. "*A degree of static electricity is that quantity which is just able to decompose nine grains of water.* 2. *A degree of current electricity is the same amount propagated during each hour of time.* 3. Where both time and length of conductor are elements, as in electro-dynamics, *a degree of electric force, or of electro-momentum, is indicated by the same*

quantity (a degree of static electricity) propagated through the length of one foot in one hour of time."

Joule uses these units in the magnetic experiments recorded in this letter, in which he determines the maximum power of magnets when fully saturated, and gives rules for their construction, which are still adopted. He had, however, reached an epoch in his research, and although he continued his magnetic investigation till 30th April, 1841, without interruption, and subsequently, at intervals, throughout his life, he clearly considered that he had exhausted the subject as far as determining the limit to the economy of the electro-magnetic engine. This is shown by the following interesting paragraph which occurs in his last paper, that in which he attributes to Jacobi the discovery of the internal resistance of the electro-magnetic engine, and which contains the substance of a lecture at the Victoria Gallery, Manchester, February 16th, 1841, the first of the very few public lectures Joule ever gave.

"With my apparatus every pound of zinc consumed in a Grove's battery produced a mechanical force (friction included) equal to raise a weight of 331,400 lbs. to the height of 1 foot, when the revolving magnets were moving at 8 feet per second. Now the duty of the best Cornish steam engine is about 1,500,000 lbs. raised to the height of 1 foot by the combustion of one pound of coal, which is nearly equal to 5 times the extreme duty that I was able to obtain from my electro-magnetic engine by the consumption of a lb. of zinc. This consumption is so unfavourable that I confess I almost despair of the success of electro-magnetic attractions as an economical source of power, for although my machine is by no means perfect, I do not see how the arrangement of its parts could be improved so as to make the duty per 1 lb. of zinc superior to the duty of the best steam engines per lb. of coal; and even if this were attained, the expense of the zinc and exciting fluids of the battery is so great when compared with the price of coal as to prevent the ordinary electro-magnetic engine being useful for any but very peculiar purposes."

Joule was then just 22 years of age. He had been working intensely for three full years, devoting his whole time to the research while living in his father's house. As regards his object in starting this research, he had met with complete failure. His papers had been obscurely published as letters in Sturgeon's "*Annals of Electricity*," and do not seem to have caught the attention of anyone capable of appreciating their merit, even if they have had a reader. But he shows no sign of flagging. He is carried on by his interest in his work and the knowledge he is acquiring. He has made himself master of the subject of magnetism, and is in many respects in front of anyone. He has mastered the difficulties of measuring 'work,' and electricity in absolute units; and has not only proved his weapons, but tasted the excitement and delight of actual battle, passing the border hitherto trodden. His mind, too, had been filled with philosophical thoughts by the laws he had discovered, while his curiosity had been excited by the unexplained incidents of his experiment. He is at no loss how to proceed, for after measuring his currents, and realizing the induced resistance of his machines, and while yet occupied in clearing up the outstanding questions, such as the limits imposed by the saturation of the magnets, he has already started a new research, and, this time, with a purely philosophical object.

CHAPTER IV.

SECOND RESEARCH.—*Communicates Results to the Royal Society.—Joule's Motive.—Attributes Proportionality of Chemical and Mechanical Effects to their Respective Quantitative Relations to the Electric Action.—Heat in Metallic Conductors.—Proportional to the Square of the Current.—Absolute Measures of Heat, Current, and Electromotive Force.—Heat Equivalent of Electrical Effect.—Heat Developed During Electrolysis.—Electric Origin of Heat.—Heats of Combustion and Electrolysis. —Intensity of Chemical Affinity of Combustibles.— Permanent and Transitory Voltaic Intensity.—Dependence of Affinity on Gaseous or Liquid States.—First Paper Before This Society.—The British Association.— Joint Research with Scoresby.—Heat Evolved During Electrolysis of Water.—Summary of Results.—Approaches Generalization.—Commences Third Research.*

Towards the end of 1840 Joule presented a paper to the Royal Society "On the Production of Heat by Voltaic Electricity," which was read in abstract December the 17th, 1840. This paper, from whatever point it is now viewed, shows throughout the hand of a master; but the matter it contained was so far in advance of the knowledge of the time, that the Royal Society declined to publish it in full in the *Philosophical Transactions*—a circumstance which not only takes a principal place in Joules' career, but has also secured a place in the history of the Royal Society. This paper was published in the *Philosophical Magazine* early in 1841, with a slightly extended title.

Joule makes no explicit statement as to what started him on this research, but this may be easily inferred. He had started his magnetic experiments, and continued them for nearly three years, under the impression that the resistance encountered by the electrical current in the wire on his magnets was simply the ordinary resistance of the wire, and was therefore the same, whether the engine was stationary or working. His philosophical attention had been directed solely to the arrangement of the electro-magnets, taking the phenomena of the battery and current for granted, as things familiar to him, which so far did not excite his philosophical curiosity. When, however, he came to measure the current, discovering that it was subject to an increased resistance when the engine was working, and further, that the electromotive force necessary to overcome this increased resistance was proportional to the work done by the engine (things he was not previously familiar with), his curiosity became excited by the phenomena of the battery and current. To satisfy this, he studied the researches of Faraday (then just published in a collected form), which made him acquainted with the works of others, which he also studied. This reading, which brought to his knowledge many facts that he was not previously acquainted with, also further excited his curiosity by revealing to him the incomplete state of knowledge as to the quantitative relation amongst the various physical phenomena, and particularly with regard to the effect of a battery in producing heat in the conductors. He had discovered for himself that the mechanical effect produced by his engine was proportional to the electric action, in Oersted's measure, expended in its production, and that the chemical action was also proportional to the electric action. There were two possible inferences to be drawn from these proportionalities. It might be inferred, as Joule was subsequently led to infer, that the chemical effect was convertible into mechanical effect in a definite ratio, and thus had a definite mechanical equivalent; the electric action being the agent of conversion. On the other hand, it might be inferred that without being convertible, or being in any way primarily related, the chemical effect and the mechanical effect were each quantitatively related to the electric action. This second inference was that which Joule at first drew. He does not explicitly say so, but it may be clearly inferred from the subsequent line of his work that the discovery of these proportionalities suggested to him the existence of definite equivalents amongst all the effects, resulting in or produced by a definite amount of electric action, as a consequence of the existence of quantitative relations between the several effects and the electric action.

In order to test these views he first investigates the heat produced by a current, and establishes, for the first time, that in a stationary conductor the amount of heat produced by a given current is proportional to the square of the current. He then proceeds to determine the quantitative relations between the heat produced, the current, and the resistance of his conductor. Taking for his units the heat required to raise 2lb. of water one degree Fahrenheit, the current necessary to effect the separation of 9 grains of water in one hour, and the resistance of a definite piece of copper wire, which he takes as his standard throughout this paper, and subsequently converts into an absolute measure by determining the current which passed through the wire, 2·26, under the electromotive force sufficient to electrolyze water. With these units he finds that the heat generated by a current in stationary metal conductors is given by the product of the co-efficient 2·14 multiplied by the electric action, or of the current multiplied by the electromotive force and the co-efficient 2·14.

He then proceeds, by the aid of this heat equivalent of electric action, to determine "The Heat developed during Electrolysis." To determine the resistance of the pairs he completes the circuit with various conductors of different resistances which he has carefully deduced, and then, by measuring the currents caused, he deduces, by "Ohm's law," the resistance of the pair in his standard units. He then measures the heat developed by measuring the capacity for heat, and the rise of temperature of the cell, and correcting for radiation, conduction and the heat generated by the dissolution of oxides, finds that the heat produced by the resistance of the pair follows the same law and quantitative relations as that produced in metallic conductors.

He then investigates the reverse process, and determines the heat produced by the passage of electricity through a decomposing cell, and compares this with the "*resistance to conduction*" in such cells, as distinct from the electromotive force necessary to cause electrolization—"resistance to electrolyzation"—in the cells. He has, therefore, first to determine this "resistance to electrolyzation." This he does by determining the part of his battery of 20 zinc-iron pairs which is necessary to overcome this resistance ($3\frac{1}{3}$ pairs), and then, as before, by the aid of conductors of known resistance, deduces by "Ohm's law" the resistance to conduction of the decomposing cell. He then finds that, as in the battery, so in the decomposing cell, the heat produced by resistance to conduction follows the same law as in metal conductors.

Having described the research which established these remarkable relations, which he has clearly been seeking, he formulates as definite conclusions—

"1st. *That if the electrodes of a galvanic pair of given intensity be connected by any simple conducting body, the total voltaic heat generated by the entire circuit (always provided that no local action occurs in the pair) will, whatever may be the resistance to conduction, be proportional to the number of atoms (whether of water or zinc) concerned in generating that current.*"

"2nd. *That the total voltaic heat which is produced by any pair is directly proportional to its intensity, and the number of atoms which are electrolized in it,*" i.e., electrical action.

"3rd. *That when any voltaic arrangement, whether simple or compound passes a current of electricity through any substance, whether an electrolyte or not, the voltaic heat which is generated in any time is proportional to the number of atoms which are electrolized in each cell of the circuit multiplied by the virtual intensity of the battery.*"

He then proceeds—" Berzelius thinks that the light and heat of combustion are produced by the discharge of electricity between the combustible and the oxygen with which it is in the act of combination; and I am of opinion that the heat arising from this and some other chemical processes is the consequence of resistance to electric conduction. My experiments on the combustion of zinc turnings in oxygen (which when sufficiently complete I hope to make public) strongly confirm this view, and the quantity of heat which Crawford produced by exploding a mixture of hydrogen and oxygen, may be considered as almost decisive of the question. In his unexceptionable experiments one grain of hydrogen produced heat sufficient to raise 1lb. of water 9·6 degrees." " Now we know from experiment 5, that the heat generated in one of Mr. Groves' pairs by the Electrolysis . . . . . of an equivalent or 32 grains of zinc . . . . . when reduced to the capacity of 1lb. of water is 9°·9. But from the table of intensities of voltaic arrangements, the intensity of Mr. Groves' pair, compared with the affinity of hydrogen for oxygen, is as 1 to 0·93, whence we have $9°·9 \times 0·93 = 9°·2$, the heat which should be generated by combustion of one grain of hydrogen, according to the doctrine of resistances."

Had this paper been published in 1840, with the guarantee of the Royal Society, it is impossible to doubt that it would have caught the attention of some of the numerous philosophers who were at the time studying these subjects. And had the facts it reveals become generally known at that time, the effect on the course of Joule's subsequent discoveries would in all probability have been great. The revelations that the heat developed by the union of two chemical elements effected in the battery is the same as that developed by combustion, and that the heat has a definite equivalent in the electromotive force between these elements, are so pregnant with suggestions, that had others entered this field of enquiry, Joule's attention might well have been diverted from the line it subsequently followed, and the completion of his work taken out of his hands. As it was, however, Joule was left to follow out his work as his own curiosity led him.

At this time he has no idea where it is likely to lead him. His philosophy has been directed to the relation between electricity and heat, just, as in his previous work, it was to the relation between electricity and work. But the fundamental character of heat has engaged his attention no more than has that of 'work'; the skin over his curiosity, arising from commonplace familiarity with these, not having yet been pricked.

For the time his attention is entirely taken up with verifying the remarkable relations he has discovered between the heat of combustion and that resulting from electrolysis. In order to establish the equality of the heats resulting from combustion and electrolysis, he first investigates, in a remarkable research, the intensity of the affinities of the various combustibles for oxygen, according to Davy's method, by using the measure of these affinities which is afforded by the electric current. To do this he has to distinguish between the permanent intensities of the electrolytic cells and those which are transitory. He first grapples with these transitory intensities, and investigates their sources; discovering that they are to a great extent owing to the condensation of oxygen on the plates during exposure to the air. He is thus led to realize that the intensity at the instant of immersion, when the negative plate is covered with a film of oxygen, represents the intensity of the affinity of the positive plate for oxygen in the non-gaseous state; while the permanent intensity represents the excess of the intensity of the affinity of oxygen for the positive plate over its affinity for gaseous hydrogen. He is also led to recognise that the affinity of one substance for another depends on the state (gaseous or liquid) of these substances.

Taking these things into account, he determines by the

electromotive force necessary to effect decomposition "the intensities of the affinities which unite gaseous oxygen with zinc, iron, potassium, and gaseous hydrogen." He then, by a method of his own, determines the absolute quantities of heat which are generated by the combustion of these bodies in oxygen, obtaining results which prove clearly " that the quantities *of heat which are evolved by the combustion of the equivalents of bodies are proportional to the intensities of their affinities for oxygen.*" The same law he had previously proved for the heat developed by electrolysis.

He then compares the actual quantities of heat produced by combustion with the quantities of heat equivalent to the electric actions in the electrolysis, finding that the latter exceed the former by one quarter. This discrepancy he explains as probably due to loss of heat in his experiments on combustion, and concludes: " I conceive, therefore, that I have proved in a satisfactory manner that the heat of combustion (at least when it terminates in the formation of an electrolyte) is occasioned by resistance to the electricity which passes between oxygen and the combustible at the moment of the union. The amount of this resistance, as well as the manner of its opposition, is immaterial both in theory and in experiment, and if the resistance to conduction be great (as it most probably is when potassium is slowly converted into potassa by the action of a mixture of oxygen and common air) or little (as it probably is when a mixture of oxygen and hydrogen is exploded), still the quantity of heat evolved remains proportional to the number of equivalents which have been consumed, and the intensity of their affinity for gaseous oxygen."

FIRST COMMUNICATION TO THE SOCIETY.

These results were communicated to this Society in a paper—"Electric Origin of the Heat of Combustion"—read 2nd November, 1841. Of this meeting Mr. B. St. J. B. Joule's diary contains the following note: " I accompanied James to hear his first paper before the Literary and Philosophical Society—Rev. J. J. Taylor in the chair; Dalton was present, and for the first time in his life moved the thanks of the meeting (and G. W. Wood seconded) to the author of the paper."

After reading this paper, which was published in the *Philosophical Magazine*, Joule was elected a member of this Society, 25th January, 1842.

At the meeting of the British Association, in Manchester, 22nd June, 1842 (at which Joule first meets Scoresby) Joule reads another paper on the same subject, in which he describes further results. The uniform discrepancy of one quarter found in his first determination between the heats of combustion and those of electrolysis he had explained as probably due to loss of heat in his experiments on combustion. This explanation is interesting as containing the first intimation of any doubt he has ever cast on any of his experiments. It is also interesting to find that his doubt was wrong, and could only have been momentary, for in this paper before the British Association he compares his own results with those of Dulong which are in close agreement, so that the previously observed discrepancies were not owing to error in his measurement of heat. Satisfied of this he has sought, in the meantime, for the cause of the discrepancies in some hitherto unrecognised actions in the cell, and has discovered a discrepancy between the electromotive force engaged in electrolysing the compound bodies and that actually used in decomposing the elements. The former including the force necessary to overcome the joint affinities of the oxide for the acid and the acid for water, as well as the affinity of the oxygen for the metal, so that while the total heat in the cell is as the product of the total electromotive force, multiplied by the current, the heat due to decomposing the electrolyte is the product of the electromotive force to overcome the last affinity multiplied by the current.

He then adopts a method of eliminating the effect on the heat of the two former affinities, when he finds the corrected theoretical quantities of heat, that is the quantities equivalent to the electric action, agree very closely with the experimental heats of combustion, in every case except that of hydrogen, where there is still a discrepancy of some five per cent., which now he attributes to still undetermined actions in the cell.

In concluding his paper, he says, " I conceive that the correctness of the idea, entertained, I believe, by Davy, and afterwards more explicitly mentioned by Berzelius, that the heat of combustion is an electrical phenomenon is now sufficiently evident. I have shown that the heat arises from the resistance to the conduction of electricity between the atoms of combustibles and oxygen at the moment of their union. Of the nature of this resistance we are still ignorant."

This remark, in which Joule clearly over-states his case, shows that at this date, June, 1842, he had not conceived the idea of the heat developed by chemical union being a measure of the mechanical potency of chemical separation, and thus being the same by whatever agency it is developed. The remark is also interesting as containing an expression of a conviction which was not sanctioned by his scientific instinct. That he is not satisfied with his proofs is shown in his next paper.

Although Joule was already engaged with Dr. Scoresby in a further research on the mechanical powers of electro-magnetism by means of Scoresby's great battery, for which he had been staying at Bradford during July, 1842, yet he found time to conduct his now world-famous investigation "On the Heat Evolved During the Electrolysis of Water," which he read before this Society on January 24, 1843. This paper also contains a summary of the results which he has so far obtained, which shows that he was already in the throes of generalization.

The first part of this research covers much the same ground as the previous.

This gives him 1·35 Daniell's cells as the electromotive force necessary to overcome the resistance to the separation of water into its gaseous elements alone. He then remarks "1·35 will very nearly represent the intensity or electromotive force required for the separation of the elements of water, and the assumption by them of the gaseous state. By these means heat becomes latent and a reaction on the intensity of the battery takes place without the evolution of free heat. It is most interesting to enquire what part of the whole intensity is due to each action."

He then describes other experiments from which he deduces that the electromotive force of 1· Daniell's cell is sufficient to separate the elements of water and give hydrogen the gaseous state, while an electromotive force of 1·45 Daniell's cells is necessary to separate the elements, and give to the oxygen as well as the hydrogen the gaseous state, whence subtracting he arrives at 0·45 of a Daniell's cell, as necessary alone to give oxygen the gaseous state, upon which he remarks—"0·45 resistance to electrolysis is equal to 2°·76." (Fah. in 1lb. water). "It would be curious to ascertain whether the same amount of caloric would be evolved by the mechanical condensation of eight grains of oxygen gas."

This is only a passing remark, but it is interesting as showing that the significance of the discovery that part of the electromotive force of his battery is occupied in performing an operation which may be reversed by mechanical means, has not escaped his notice, and has excited his curiosity. This comes out again in the observations he makes at the end of his paper. There are six of these observations. In the 4th he says :—

"Faraday has shown that the *quantity* of current electricity depends upon the number of chemical equivalents which suffer electrolysis in each cell, and that the intensity depends on the sum of the chemical affinities. Now both the mechanical and heating powers of a current are, per equivalent of electrolysis in any one of the battery-cells, proportional to its intensity or electromotive force. Therefore, the mechanical and heating powers of a current are proportional to each other."

In the 5th observation, he says :—

"The magnetic electrical machine enables us to convert mechanical power into heat by means of the electric currents which are induced by it, and I have but little doubt that by interposing an electro-magnetic engine in the circuit of a battery a diminution of the heat evolved per equivalent of chemical change would be the consequence, and this in proportion to the mechanical power obtained." And in a note (Feb. 18, 1843), "I am preparing experiments to test the accuracy of this proposition."

Had he stopped here, it would have been almost impossible to avoid the conclusion that he had not only realized the equivalence of the mechanical and heating powers of an electric current, but that he had already realized the probable equivalence of the mechanical and heat effects. Nor does his sixth observation contradict this conclusion. In this, he says:—

"Electricity may be regarded as the grand agent for carrying, arranging, and converting chemical heat," and illustrates this by referring to the definite amounts of heat which would otherwise be developed in the cells of the battery, which are, by the current, transferred by the conductors and developed in a decomposing cell, or in a resistance coil.

In his seventh observation, however, he reasserted his opinion that he has "put the beautiful theory of chemical heat, first suggested by Davy and Berzelius, beyond all question," showing that he had not yet realized the primary character of the relation between heat and mechanical effect, which might be deduced from the results he already had obtained, and which he was then on the verge of discovering. Nor is this all. So far throughout his work he has given no hint that he has ever considered the character of heat, whether a "substance" or a mechanical phenomenon, neither has he, in the least, described the mechanical significance of "work." Of course, he might have considered them all the same; but the proof that he had not done so is furnished in a note he added to this paper on February 20, 1844, after his discovery was complete.

That the three fundamental quantities, work, heat, and electricity, with which Joule had been occupied all this time should have been accepted by him as having a natural right to exist without further question as to their origin, receives its explanation in the difficulty always found in fixing attention on what is familiar. Heat and work in no way excited his curiosity. It was the rarer and more occult phenomena of electro-magnetism which first caught his attention, leading him first to pay attention to electricity, thence to heat, and now to "work."

In the note already mentioned, he says that he had always been "strongly attached to the theory which regards heat as motion amongst the particles of matter." This must have been an instinctive attachment, formed without question or consideration, for a theory of infinite possibility resting on a rational base, as against a dogma opposed to many familiar facts.

Joule had now discovered and described all the equivalences but one on which the conservation of energy is founded; the heat, and the chemical equivalents of electrical effect, and the heat equivalents of chemical effects. He has not yet generalized, because he has not realized, that the underlying principle is mechanical effect. But by January 14th, 1843, his attention had been arrested by the mechanical character of the effect of that portion of the electromotive force used in electrolysis, which renders the elements gaseous and still more strongly by the constant ratios between the mechanical and heating powers of a current and between these powers and the electromotive force. He looks for further developments, but apparently without, as yet, realizing their true character or importance.

CHAPTER V.

THIRD RESEARCH.—*Heat Generated or Transferred.—Arrangement not Generation of Heat in Voltaic Apparatus.—The Heat Developed in the Entire Circuit by Magneto-Electricity not the Result of Arrangement.—Discovers Heat in the Revolving Armature.—Determines Relation between Heat and Electric-Action; the Same as with Voltaic Current.—Generation and Destruction of Heat by Mechanical Means by the Agency of Magneto-Electricity.—Constant Ratio between the Heat and the Power.—First Determination of the Mechanical Equivalent of Heat.—The Climax of Joule's Researches.—Attends British Association at Cork, 1843.—Dazzled by Possibility of Practical Results.—Conclusions not Altogether Justified.—Postscript.—Mechanical Effect Converted into Heat by Friction.—Law of Conservation Realized.*

In his paper, read before the Literary and Philosophical Society, January 24th, 1843, in his fifth observation Joule had said:—" The magnetic electrical machine enables us to convert mechanical power into heat by means of the currents which are induced by it." Further consideration, however, seems to have suggested to him that this point was still lacking experimental proof; for his next research is, in the first place, obviously and avowedly directed to ascertain whether the heat which is rendered apparent in one part of the circuit of the armature of the machine was *generated* by the current, or merely *transferred from the coils* in which the magneto electricity was induced, the coils themselves becoming cold. Transference of heat played such an important part in the attempts which had been made to reconcile the dogma that caloric was a material with observed phenomenon, that there would have been nothing remarkable at that time in Joule, although for the moment carried forward by his own experiments and physical instincts, being brought up again by the physical casuistry of his time. This, however, does not appear to have been the case, at least not altogether. It is his own discovery recapitulated, in his sixth observation, that the heat evolved in the circuit of the voltaic battery is the definite chemical heat due to the union of the elements in the cells transferred to other parts of his circuit, which suggests, by analogy, that the heat developed in one part of the circuit of the magnetic engine *may* be merely transferred from another.

There is more interest in this stage of Joule's work than is apt to appear on the surface. So far he has shown no indication of holding any view heterodox to those generally accepted at the time respecting heat. He has shown that the heat developed in the voltaic circuit exactly corresponds with the heat developed by the combustion of the same elements, and offered this as proof that the heat of combustion is developed by electricity. But in this there is nothing at variance with the then accepted hypothesis that the heat developed by the union of the elements, however brought about, was latent in the elements before union, which hypothesis, therefore, so far explains the source of the heat in the voltaic circuit. And then he had established facts which "might seem to prove that *arrangement* only, not *generation* of heat, takes place in the voltaic apparatus, the simple conducting parts evolving that which was previously latent in the battery." In the generation of magneto-electricity there is no chemical union to supply the heat, nor any change in the nature of parts of the machine, so that if more heat is evolved in one part than is abstracted from another this heat has for its only source the effect of the magneto-electric current. But in order to ascertain if this were the case it was necessary to take account of the changes of temperature of each part of the machine. This was the investigation Joule now undertook with his mind perfectly open. "I resolved," he says, "to clear up the uncertainty with respect to magneto-electrical heat."

The research was one of extreme difficulty, and as an experimental feat probably surpasses anything ever accomplished. Joule confined his attention to the action of the magneto-electric current caused in the coil of a revolving armature, or "small compound electro-magnet enclosed, coil and all, in a revolving vessel of water." From the vessel the ends of the coil protruded, and enabled him, by means of a mercury commutator, to include a galvanometer of small resistance in the circuit. He then measured the current of magneto-electricity generated and the heat developed in the coil—as shown by the rise of temperature of the water—when the armature revolved between the poles of a powerful electro-magnet. He took elaborate precautions to prevent loss of heat by radiation and air currents; notwithstanding which, he says, "I was led into error by my first experiment; the water had lost heat even when the temperature of the room was such as led me to anticipate a contrary result." "I did not stop," he continues, "to inquire into the cause of the anomaly, but provided effectually against its interference with the subsequent results by alternating the experiments with others made under the same circumstances, except as regards the communication of the battery with the stationary magnets which was in these instances broken."

The result of these alternate experiments, was that with the magnets excited, the temperature of the water surrounding the moving electro-magnet rose one-twentieth of a degree Fahrenheit in 15 minutes, while with the contact broken, it fell by the same amount, giving him one-tenth of a degree Fahrenheit rise in temperature, as the heating effect induced in the revolving magnet and circuit by the stationary magnets.

"Having thus detected the evolution of heat from the coil of the magneto-electrical apparatus, my next business was to confirm the fact by exposing the revolving electro-magnet to more powerful magnetic influences."

This he does, using stationary electro-magnets of different power, also stationary steel magnets; he thus obtains results ranging from one-tenth degree Fahrenheit to two degrees and a half. Having thus proved that heat is generated by the magneto-electric current, he puts a finish to this research, first by determining the heat evolved in the iron core when revolving without its coil between the poles of the electro-magnet; and substracting this obtains the relation between the heat generated in the coil of his revolving magnet and the product of the square of the current, multiplied by the resistance of the coil, so as to see how far it corresponds with the heat that would be generated by a voltaic current. Whence he finds, making allowance for the intermittent character of the magneto-electric current, "That *the heat evolved by the coil of the magneto-electrical machine is governed by the same laws as those which regulate the heat evolved by the voltaic apparatus, and exists also in the same quantity under comparable circumstances."*

Having completely settled these preliminary questions he now reverts to the prediction he had made in his 5th observation, January 24, 1843, viz., " I have little doubt that by interposing an electro-magnetic engine in the circuit of a battery a diminution of the chemical heat evolved per equivalent of chemical change would be the consequence and this in proportion to the mechanical power obtained."

"For this purpose it was only necessary to introduce a battery into the magneto-electrical circuit, then by turning the wheel in one direction I could oppose the voltaic current, or by turning in the other direction I could increase the intensity of the voltaic by the assistance of the magneto-electricity." "In the former case the apparatus possessed all the properties of the electro-magnetic engine, in the latter it presented the reverse, namely, the expenditure of mechanical power."

As the result of these experiments he finds that the heat developed is strictly in accordance with the laws he has previously discovered, namely, proportional to the product of the square of the current and the resistance of the circuit, "and is not affected either by the assistance or resistance which the magneto-electricity presents to the voltaic current." That is to say, a definite current in the circuit of the armature (or moving electro-magnet) causes the same heat, whether the armature is at rest or in motion— whether the current is caused (1) by the electromotive force of the battery alone, (2) by the magnetic induction on the moving magnet, (3) or by both of these acting in conjunction or opposition.

In the first case, the heat produced in the circuit corresponds to the heat of the chemical effects in the battery—is merely the latent heat of the elements combined transferred to the coil; in the second case, a similar amount of heat is generated without any chemical action, solely by the "work" done in moving the machine. In the third case, the same amount of heat is developed in the coil, but the chemical action in the battery is increased or diminished in exact proportion to the electro-motive force, with or against the current, induced by the motion of the armature. So that, "the heat due to a given chemical action is subject to an increase or diminution directly proportional to the intensity of the magneto-electricity assisting or opposing the voltaic current." "*We have, therefore, in magnetic-electricity, an agent, capable by simple mechanical means of destroying or generating heat.*" Having thus verified the first part of his prediction, he now turns to the second part.

He says: "Having proved that heat is generated by the magneto-electrical machine, and that by means of the inductive power of magnetism we can diminish or increase at pleasure the heat due to chemical changes, it became an object of great interest to inquire whether a constant ratio existed between it and the mechanical power gained and lost. For this purpose, it was necessary to repeat some of the previous experiments, and to ascertain, whether at the same time, that mechanical force was necessary in order to turn the apparatus."

His previous experience in measuring the power of his electro-magnetic engines now stands him in good stead, and he makes thirteen experiments, driving the apparatus as a magneto-electric machine, using it as an electro-magnetic engine, varying the intensity of his magnets and

MECHANICAL VALUE OF HEAT.   65

the resistance. The results he obtains give for the work necessary to raise 1lb. of water 1 degree Fahrenheit, quantities which lie between 1,026 and 587 lbs. raised one foot. Upon which he says:—"The foregoing are all the experiments I have hitherto made on the mechanical value of heat. I admit there is considerable difference between some of the results, but not, I think, greater than may be referred with propriety to mere errors of experiment. I intend to repeat the experiments with more powerful and more delicate apparatus. At present we shall adopt the mean result of the thirteen experiments given in this paper, and state generally that"

"*The quantity of heat capable of increasing the temperature of a pound of water one degree of Fahrenheit's scale is equal to, and may be converted into, a mechanical force, capable of raising 838lbs., to a perpendicular height of one foot.*"

This result is the climax of Joule's researches. For although he was immediately able, by using simpler means, to attain greater accuracy in the determination of the mechanical equivalent of heat; this was of small importance compared with the philosophical insight which resulted from the comprehensiveness and sequence of his investigation.

Commencing with the electro-magnetic engine, and discovering the measure of the electric action equivalent to the work done, then showing that the heat developed in a conductor has also an equivalent in the same measure of electric action, and further that the heat developed in the cells of a battery has an equivalent in the same action when proper allowance is made for secondary actions; and so discovering, by means of this measure of electrical action,

F

that the chemical action in the cells of the battery, as expressed by the product of the chemical affinity of the elements multiplied by the quantity of equivalents electrolyzed, is equivalent not only to the electrical action but to the heat developed in the entire circuit. Thence showing by absolute measurement that the heat so produced corresponds with the heat produced by the combustion of the same elements, and further, that it corresponds with the heat equivalent of the electric action used in the electrolytic decomposition of the electrolyte formed by combustion provided the elements are brought back to their initial gaseous state. Then the determination of the electric action necessary to convert liquid oxygen into gaseous oxygen, and the determination of the equivalent heat.

Finally proving, still by aid of the measure of electric action, that the heat developed by "work" done on the magneto-electro machine, and that the heat equivalent of the electric action spent in doing work in the electro-magnetic engine, have a constant equivalent in the 'work' of about 838 lbs. raised 1 foot for each unit of heat which would raise 1 lb. of water 1 degree Fahrenheit.

Joule had thus, by means of the measure of electric action, traced a definite quantity of physical effect throughout the whole region of physics, recognising it in all the transformations it was capable of undergoing, and discovering all its modes, flinching at no experimental difficulties to keep it in view until he had brought it again into full light as work or mechanical energy.

Each one of these discoveries constitutes in itself a great step in physical science, and would have established the reputation of the discoverer; but the flood of light thrown by their accumulative effect over the whole region of physics was such as to obscure their individual importance, and merge them all in the one discovery of the mechanical equivalent of heat, which, although it was the crowning discovery, and of the greatest general importance, besides being the most easily intelligible, was standing by itself, only one step towards the end which had been reached.

For the exaltation of the "mechanical equivalent of heat" over his other discoveries, Joule was himself in some degree responsible. Although the accounts of his earlier discoveries were scattered about in various publications, and some of them not then fully published, yet in writing the account of the first determination of the mechanical equivalent of heat, not only does he make no attempt to bring them together, but he does not even refer to their general significance.

This research had been conducted in the six months between January and August, 1843. On August 15th Joule starts, in company with Mr. Eaton Hodgkinson, for Cork, where the British Association was to assemble on August 17th. His paper "On the Calorific Effects of Magneto-Electricity and the Mechanical Value of Heat" was read before the Chemical Section, August 21st, and was received in general silence. The paper is dated July 23rd, 1843, and was published in full in the *Philosophical Magazine*.

In the paper itself, according to his wont, Joule confines himself to very few remarks upon his work, but it, nevertheless, contains inherent evidence, that, at the time of writing, the philosophical significance of his discoveries was rendered obscure to him by the dazzling effect of the, to some extent meretricious, practical conclusions which he conceives "may be drawn from the convertibility of heat and mechanical power into one another, according to the above absolute numerical relations."

The chief of these conclusions, although implied rather than explicitly stated in the paper, but explicitly stated four years later, is that "In the case of the steam-engine, by ascertaining the quantity of heat produced by the combustion of coal, we can find out how much of it is converted into mechanical power, and thus come to the conclusion as to how far the steam-engine is susceptible of further improvements. Calculations made upon this principle have shown that at least ten times as much power might be produced as is now obtained by the combustion of coal." There is nothing absolutely wrong in this conclusion, but it ignores the physical impossibility of the conditions necessary to such realization; conditions not by any means understood at the time, nor till six years after, and then as a consequence of Joule's discovery. Yet, accepted in its simplicity, this conclusion was at variance with all experience obtained throughout the career of the steam engine.

That, accepted in its simplicity, this conclusion should have dazzled Joule for the time is not to be wondered at; and that so accepting it, he should have been led to call the mechanical equivalent of heat, which he had discovered, "the mechanical *value* of heat," and use this as the title of his paper, is not a matter of surprise. It was, nevertheless, unfortunate after he had throughout his work been consistently seeking equivalences, and on finding the crowning equivalence, had not only treated it, but also considered it throughout his investigation as an equivalence, that he should have adopted another, and at least a questionable expression, in the titles of his papers.

CONCLUSIONS NOT ALTOGETHER JUSTIFIED.

There is, too, another questionable statement in Joule's paper of July, 1843, which shows that he was somewhat dazzled and confused in the general view he took of the results he had obtained—a confusion directly connected with the practical conclusions which he draws. This is in the general statement of the results already quoted. He therein says: "The quantity of heat capable of raising the temperature of a pound of water one degree Fahrenheit's scale is equal to and *may be converted into* a mechanical force, capable of raising 838 lbs. to a perpendicular height of one foot." It is as to the expression *may be converted into*, that the important question arises—is this established by his research? Has he, as yet, shown that heat may be converted into power? He has unquestionably shown that power may be converted into heat, and that electrical action and chemical action, by means of electrical action, can be converted into their equivalents of either power or heat, but in the detail of his papers he has hitherto been careful to speak of the production of power by electrical action, "as attended by a diminution of the *free* heat that would otherwise have appeared in the circuit"—not in any case implying that he has converted free heat into mechanical power any more than into chemical action, necessary to effect electrolytic separation in the decomposing cell. Thus, if the term heat in his general statement is understood to mean free heat, this statement as to the conversion of heat into work was at the time unproved. That Joule considered the statement true in this sense is shown by the practical conclusions already discussed, but in making the statement he no doubt referred to the *latent-heat* residing in the battery or the uncombined elements, which he had discussed in a previous paper, and in this sense the statement was justified; for the time, however, it is clear that the essential distinction between these two heats escaped his notice, and allowed him to overstep the real significance of his results.

It may seem unnecessary to call such pointed attention to misconceptions that were subsequently corrected by Joule himself; but in their influence on Joule's subsequent career they were not unimportant, as they tended for the time to render his results apparently inconsistent with observations and laws as to the performance of work by means of heat, which were then exciting an increasing interest, and so led some of those best able to judge to reject Joule's results at first sight, without ascertaining their real significance.

Although Joule adhered to his practical conclusion as to the possibility of the conversion of heat into work, for five or six years, its dazzling effect in obscuring the philosophical significance of his results was only momentary. Having written his paper, he seems at once to have perceived the bearing of his discoveries on phenomena, which had not hitherto attracted his attention. It is at least remarkable that throughout his papers so far he has not made any reference to the heat produced by friction, although he frequently mentions that part or the whole power developed in his engines was expended in overcoming friction. The papers themselves would bear the interpretation that he had not hitherto realized that there was any other agency but that of electricity, by which heat and work were convertible; a view which is to a great extent borne out by the remarkable manner in which he commences a most remarkable postscript, which contains a clear and somewhat full statement of his views at the time, showing, in somewhat crude language, that he had already, in August, realized the law now called the Conservation of Energy as the result of his discoveries. This postscript, which was added in August to his paper of the 23rd of July, is as follows :—

P.S.—" We shall be obliged to admit that Count Rumford was right in attributing the heat evolved by boring a cannon to friction and not (in any considerable degree) to any change in the capacity of the metal. I have proved experimentally that *heat* is evolved by the passage of water through narrow tubes. My apparatus consisted of a piston perforated by a number of small holes working in a cylindrical glass jar containing 7 lb. of water. I thus obtained one degree of heat per lb. of water from a mechanical force capable of raising about 770 lb. to the height of one foot, a result which will be allowed to be very strongly confirmatory of our previous deductions. I shall lose no time in repeating and extending these experiments, being satisfied that the grand agents of nature are by the Creator's fiat indestructible, and that whatever mechanical force is expended an exact equivalent of heat is always obtained.

"On conversing a few days ago with my friend, Mr. John Davies, he told me that he had himself, a few years ago, attempted to account for that part of animal heat which Crawford's theory had left unexplained by the friction of the blood in the veins and arteries, but that finding a similar hypothesis in Haller's " Physiology " he had not pursued the subject further. It is unquestionable that heat is produced by such friction, but it must be understood that the mechanical force expended in the friction is a part of the cause of the affinity which causes the venous blood to unite with oxygen ; so that the whole heat of the system must still be referred to chemical changes. But if the animal were engaged in turning a piece of machinery, or in ascending a mountain, I apprehend that in proportion to the muscular effort put forth for the purpose a diminution of the heat evolved in the system by a given chemical change would be experienced.

"I will observe, in conclusion, that the experiments detailed in the present paper do not militate against, though they certainly somewhat modify, the views I have previously entertained respecting the electrical origin of chemical heat. I had before endeavoured to prove that when two atoms combine together, the heat evolved is exactly that which would have been evolved by the electrical current due to the chemical action taking place, and is therefore proportional to the intensity of the chemical force causing the atoms to combine. I now venture to state more explicitly that it is not precisely the attraction of affinity, but rather the mechanical force expended by the atoms in falling towards one another which determines the intensity of the current, and, consequently, the quantity of heat evolved, so that we have a simple hypothesis by which we may explain why heat is evolved so freely in the combination of gases, and by which, indeed, we may account 'latent heat' as a mechanical power prepared for action as a watch spring is when wound up. Suppose, for the sake of illustration, that 8lb. of oxygen and 1lb. of hydrogen were presented to one another in the gaseous state and then exploded, the heat evolved would be about 1 degree Fahr. in 60,000 lbs. of water, indicating a mechanical force expended in the combination equal to a weight of about 50,000,000 lbs. raised to the height of 1 foot. Now, if the oxygen and hydrogen could be presented to each other in a liquid state, the heat of combination would be less than before, because the atoms in combination would fall through a less space. The hypothesis is, I confess, sufficiently crude at present, but I conceive that ultimately we shall be able to represent the whole phenomena of chemistry by exact numerical expressions, so as to be enabled to predict the existence and properties of new compounds.

"J. P. J.

" August, 1893."

CHAPTER VI.

EFFORTS TO CONVINCE THE SCIENTIFIC WORLD.—
*Pre-eminence in Knowledge of Physical Science.—General
Silence; the Highest Tribute to the Greatness of the
Advance.—Friendly Sympathy.—Oakfield.—Researches
in his New Laboratory.—Rarefaction and Condensation
of Air.—Difficulties. — No Latent Heat.—Convertibility
of Free Heat into " Work."—Dynamical Theory of
Heat.—Development of Davy's Dynamical Theory of
Gases.—New Theory of the Steam Engine.—Criticism
of Carnot's Theory. — Heat Discharged into the Con-
denser.—Discussion of Results.—Joule Ignores the Truth
of Carnot's Theory.—Indestructibility of Caloric Proved.—
Definite Dynamical Theory of Gases. — Imperfect
Acquaintance with Mechanical Philosophy. — Absolute
Zero of Temperature.—First Determination with the
Paddle. — Realization of Dynamical Significance of
" Work."—Extent of Experimental Work. — Research
with Scoresby; Visits to Bradford. — Essay to the
Institute of France.—Research on Effect of Magnetism
on the Dimensions of Iron and Steel.—Joint Research
with Sir Lyon Playfair; Atomic Volumes.*

In August, 1843, Joule is in his 25th year, and has
been at work on his research for four years and a half.
That he has devoted his whole time to the work, and has
had all the facilities an indulgent father could supply, must
be allowed, but still the scientific precosity he has displayed,

as well as his achievements, remain unique. In the subjects of electricity, magnetism, electrolysis, heat, electro-magnetism, and thermal-chemistry, he has mastered the knowledge which existed at the time up to its latest developments. He has himself experimented over the whole ground, at every step rendering that definite which was hitherto vague. He has discovered the definite relations between electricity, heat, chemical action, and power ; and by his philosophy he has carried his knowledge of physics to a height from which he can now look down on his contemporaries still struggling to reach higher points in the various branches of physics, but neglecting the connecting ridges which he has traversed, although he now sees these ridges all united in the summit which he has scaled.

It is not his own doing that he is so absolutely alone in possessing the knowledge. He has recorded at the time the steps by which he has proceeded. These records he has posted in the most exposed places within his reach, and has tried to do more ; and now that he has reached the summit without waiting to explore, he again does his best to reach the ears and eyes of his contemporaries, but without as yet inducing any to follow him.

The fact that these early papers of Joule were, at the time, apparently ignored by the many eminent physicists then living, though apt to inspire the present reader with a feeling of astonishment, if not of indignation, at the generation for their prejudice and neglect, was, in truth, the highest tribute that could be paid to the greatness of the advance in philosophy which he had made. The present reader is apt to forget, not only that the facts revealed in these papers, which have become familiar as accepted truths, were for the time novel, and apparently heterodox, but that the very terms in which these facts are expressed, which have become familiar as part of our language, were at that time, of necessity, only to be understood after the significance of the facts had been appreciated. It is one thing to feel satisfied with the cogency of an argument after it has the authority of general acceptance, and quite another so to appreciate this cogency for oneself as to give one confidence to avow acceptance with it in the face of general opinion.

At the time of which we are writing there were a band of physicists in this country, alone, who, by the advances they made in the various branches of physics, have rendered the era remarkable. Faraday, then engaged in completing, his experimental researches in electricity; Daniell and Grove, whose work contributed directly to Joule's results; and besides these, Graham, then in the midst of his researches on the diffusion of gases; Miller, Wheatstone, Whewell, Herschel, Forbes, Airy, Apjohn, Henry, Herapath, and others. The now classical works of these men, fundamental as they were to the various branches of science, place them in the very highest position for honesty of purpose and enthusiasm in scientific advancement; and remove them as far above the suspicion of prejudice or neglect as is possible for any human being. Those who were subsequently the first to follow Joule belonged to another generation. Of these Thomson (Sir William) and Stokes (Sir Gabriel) were still at Cambridge, Stokes having taken his degree of B.A. the year before, and Thomson being only in his third year, Rankine was still preparing for the career of an engineer, being 22 years of age, while Tate and Maxwell were still at school.

That the obscure and scattered manner in which Joule's papers were published must have hindered his work from becoming as well known as it might otherwise have been is certain ; but the fact that it called forth no comment does not show that it was not read, or if read was rejected. Two of the most important of his papers were read at the British Association and with several others published in full in the *Philosphical Magazine*. These were doubtless heard or read, and the complete silence in which they were received shows not that they were rejected, but that the views they contained were so much in advance of anything accepted at this time that no one had sufficient confidence in his own opinion, or was sufficiently sure of apprehending the full significance of the discoveries on which these views were based, to venture an expression of acceptance or rejection. The "angels feared to tread," and perhaps the most remarkable thing is that in this case there were no fools. The position, however, is not without ample precedent besides that of Kepler. And although Joule, thus compelled against his will to remain the only individual in the world possessing the knowledge of and realising the significance of this fundamental law of the universe, was, as he says, very anxious to convince the scientific world of the truth of the positions he maintained, he was in no way discouraged by the silence in which his papers were received, though this silence remained unbroken for six more years.

He was not altogether without friendly sympathy and interest. He had that of Dr. Scoresby, with whom he had been continuing his magnetic researches, as well as of his family and friends in Manchester, and particularly of the members of the Literary and Philosophical Society.

On 19th March, 1844 Joule's father removed from

Broom Hill to Oakfield, Whalley Range, where he had built Joule a laboratory, affording him greatly increased facilities for his work.

In August, 1843, he had already engaged in experiments for determining the mechanical equivalent of heat by the friction of water. These he continues, with various apparatus, as fast as the time occupied in adapting his apparatus will allow.

In the meantime, he proceeds to carry his experiments into other regions with the same object, and to collect, complete and publish notes concerning incidental phenomena that he has observed during his research. On December 26th, 1843, he read before this Society a paper "On the Intermittent Character of the Voltaic Current in Certain Cases of Electrolysis; and On the Intensity of Various Voltaic Arrangements." This, although a very important paper on the subject of electrolysis, containing definite measurements and explanations new at the time, refers chiefly to matter collected during his previous electrical researches.

His next paper, "On the Changes of Temperature Produced by the Rarefaction and Condensation of Air," read before the Royal Society, June 20th, 1844, contains an account of one of the most important, as well as most remarkable, researches Joule made for proof of the generality of the mechanical equivalent of heat and its accurate determination. It also contains the first definite development of the dynamical theory of matter and heat. This paper was rejected for publication by the Royal Society, and was published in the *Philosophical Magazine*, in May, 1845.

The relations between the heat and work in the expansion and contraction of elastic fluids are unquestionably the most important of any. This Joule recognises. In the introduction to his paper he says, "An inquiry of great interest in a practical as well as theoretical point of view, owing to its bearing upon the theory of the steam engine." The difficulties of the enquiry were, however, so great as to have been impossible to any one but Joule. And besides presenting those difficulties which Joule actually overcame, perhaps the greatest difficulty was to conceive any possible determination that would not depend on a knowledge of the properties of gases and the relations amongst these properties which were not measured or formulated till some years afterwards, and then by the aid of Joule's discoveries ; so that not the least interesting part of this research is the way in which Joule, having carefully considered the subject, and perceived the difficulties of the reasoning, and the inadequacy of the existing knowledge, was able, by keeping in view the simple fact that he wanted to determine, namely, the relation between "work" done in compressing air and heat developed, to devise experiments which would yield this relation without raising any question as to the other properties of the air. Having devised the experiments, the courage he displayed in undertaking what obviously required a degree of experimental accuracy far beyond anything previously attained, and which has never been repeated, is no less interesting. From the light they throw on this part of his research, and from other reasons, the first few paragraphs of this paper are of interest. He says :—

"Dr. Cullen and Dr. Darwin appear to have been the first who observed that the temperature of air is decreased by rarefaction and increased by condensation. Other philosophers have subsequently directed their attention to the subject. Dalton was, however, the first who succeeded in measuring the change of temperature with some degree of accuracy. By the employment of an exceedingly ingenious contrivance, that illustrious philosopher ascertained that about 50° of heat are evolved when air is compressed to half of its original bulk, and that on the other hand 50° are absorbed by a corresponding rarefaction."

This paragraph, besides showing that Joule had studied the determinations as to the thermal properties of gas then existing, is also interesting as containing the first reference which he has had occasion to make to the works of his great master. He then proceeds:—

"There is every reason for believing that Dalton's results are very near the truth, especially as they have been exactly confirmed by the experiments of Dr. Ure, with the thermometer of Breguet." "But," he continues, "our knowledge of the specific heat of elastic fluids is of such an uncertain character that we should not be justified in attempting to deduce from them the absolute quantity of heat evolved or absorbed. I have succeeded in removing this difficulty by immersing my condensing pump and receiver into a large quantity of water, so as to transfer the calorific effects to a body which is universally received as the standard of capacity." Then, after describing the pump, he proceeds : "Anticipating that the changes of temperature of the large quantity of water which was necessary to surround the pump and receiving apparatus would be very minute, I was at great pains in providing a thermometer of extreme sensibility and very great accuracy. A glass tube of narrow bore having been selected, a column of mercury, 1 inch long, was introduced, and gradually advanced so that the end of the column in one position coincided with the beginning of the column in the next. In each position the length of the column was ascertained to the one four-thousandth part of an inch, by means of an instrument invented for the purpose by Mr. Dancer."*

"Afterwards the tube was covered with a film of beeswax, and each of the previously measured spaces was divided into twenty equal parts by means of a steel point carried by the dividing instrument; it was then etched by exposure to the vapour of fluoric acid. The scale thus formed was entirely arbitrary; and as it only extended between 30° and 90° it was necessary to compare the thermometer with another constructed in the same manner but furnished with a scale including the boiling as well as the freezing point. When this was done it was found that the ten divisions of the sensible thermometer (occupying about half an inch) were nearly equal to the degree of Fahrenheit." Then comes the most remarkable statement : "Since by practice I can easily estimate with the naked eye *one-twentieth* of each of these divisions, I could with this instrument determine temperatures to the one two-hundredth part of a degree."

Considering that a skill in thermometry previously undreamt of, and never since surpassed, was the principal means by which Joule finally attained the extreme accuracy in his determination of the mechanical equivalent of heat, the foregoing passages are not without interest, followed, as they are, by the experimental determination of differences

---

* "Of the firm Abraham and Dancer, Cross-street, Manchester. I have great pleasure in acknowledging here the skill displayed by this gentleman in the construction of the different parts of my apparatus; to it I must, in a great measure, attribute whatever success has attended the experiments detailed in this paper."

G

of temperature less than the tenth part of a degree (one twentieth of an inch on Joule's thermometer), which were correct to the two-hundredth part of a degree, as is shown by the results deduced from them.

With these appliances Joule makes three series of experiments. In the first he pumps his carefully-dried air slowly into a receiver, then determines the increase of temperature in the surrounding water. This increase is, he considers, due to the friction of his pump, to radiation from the room, as well as to the work done in compressing the air. He therefore makes experiments, from which he determines the amount of the rise due to the first two causes, and subtracting this he gets the rise of temperature $0°·345$ Fah. due to the last cause. This rise, multiplied by the capacity for heat of the water, gives him the heat resulting from the condensation of the air. He has now to obtain the work which has been done in compressing the gas. This he does, not by measuring the resistance and motion of the pump, but by measuring the volume of the compressed air when expanded again to the atmospheric pressure. This is $21·654$ times the volume of his receiver, which he has calibrated. Then, by Boyle's law, he estimates that the final pressure of the air in his receiver is $21·654$ times the pressure of the atmosphere at the time, and by this means and the hyperbolic relation between the pressure and volume he estimates the work done. This process would be strictly accurate, provided the air in the pump had maintained a constant temperature, which it would not have done had not sufficient time been allowed for the heat resulting from condensation to pass from the air into the water during the intervals of compression. Joule nowhere in his paper mentions the necessity of this condition; but

SECOND SERIES ON LATENT HEAT.

from the slowness with which he worked his pump—from ten to five strokes a minute—it may be inferred that he was alive to it, and conceived that he had eliminated it.

This he seems very nearly, but not quite, to have done, for the first series of experiments gave him a mechanical force of 823 lb. falling one foot to raise 1 lb. of water one degree, and a repetition with slower pumping gave 795. Both these quantities are above the true value 772, as he finally determined it, and the difference may very well have been due to a small excess of temperature in the pump.

The second series of these experiments were made to determine the change in temperature due to change in the latent heat of the air with change of volume. He does not state his object, which, however, is clear from the experiments. Two equal receivers are connected by a pipe with a cock in it, initially closed. Into one vessel he compresses air, and the other he exhausts, then immerses the whole in water until they are at an equal temperature with the water, which temperature he carefully measures; then he opens the intervening cock and allows the air to flow from the full vessel into the empty, until they are at equal pressure. Now, from his first series of experiments, it would appear that the air, as it is compressed into the empty vessel, must rise in temperature, which, if there is no compensating action by the expansion of air in the full vessel, will raise the temperature of the water, but since the work done by the air in the full vessel is all expended on the air in the empty vessel, it would follow that an amount of heat equivalent to this work must be abstracted from the full vessel, and so that there should be an exact compensation, unless, indeed, there were some change in the latent heat. By this means he has eliminated every possible error in testing the mechanical character of the thermal changes caused by compression and rarefaction of air. The results fulfilled his anticipation most emphatically, there being no sensible change in the temperature of water.

He then varied the experiment by immersing each of the receivers in a separate vessel, and the connecting piece in a third vessel, when he found that 2·36 units of heat had been abstracted from the receiver in which the air had been expanded, 0·31 from the can in which the connecting piece was immersed, and that 2·38 had been produced in the other receiver, leaving 0·01 units of heat for what were lost in the short lengths of pipe between the cans.

These two experiments, so simple and yet so crucial, as to the convertibility of heat into work, were not easily available for determining the mechanical equivalent of heat; but they afforded such forcible evidence of the purely mechanical character of the elasticity of air, as to start immediately, in the mind of Joule, the now established dynamical theory of gases.

The third of these series of experiments was again devoted to the evaluation of the equivalent. It is, in fact, a modification of his second experiment adapted to that evaluation for which it is much better suited than his first series. Here he uses one receiver with a cock leading into a long pipe coiled up and immersed in the same calorimeter as the receiver. The open end of the pipe passes into a pneumatic trough for measuring the volume of air. On the cock being slightly opened the air passes slowly out of the receiver through the pipe and escapes at the temperature of the water doing work only against the pressure of the atmosphere. Thus the work that was done by the escaping air was simply the product of its volume, multiplied by the atmospheric pressure ; and this being the only work done by the escaping air would be the equivalent of the only heat abstracted from the receiver and the pipe, other than that which passed to and from the surrounding water. Three sets of experiments, with this contrivance, gave values for the equivalent of 820, 814, and 760, with a mean 798, which is only 3·3 per cent. above the final value, although the actual differences of temperature measured did not exceed 0·1 degree Fah. Neither are the discrepancies, which do not amount to 5 per cent., nor yet the slight excess of the mean, to be attributed solely to errors of thermometric measurement ; for it was necessary to stir the water to equalize the temperature before reading this temperature, and the work done in this stirring exercised such a sensible effect on the temperature that this effect had to be determined and deducted ; affording a source of error.

These determinations, by the first and last series of experiments, of the relations between the heat and work involved in the expansion and contraction of air would not, in themselves, have afforded conclusive evidence of the convertibility of heat into work like that deduced from Joule's magnetic or friction experiments ; for these experiments still left it possible to suppose that the heat evolved or absorbed was merely transformed from latent to sensible and *vice versâ* from sensible to latent, in consequence of the changes in the density of the gas. That Joule was fully aware of this was shown by his undertaking the second series. In this the air underwent exactly the same final change as in his third series, but without doing any external work, in which case if there was any change from sensible heat to latent it would still have appeared. It was only on finding in this second series that there was no latent heat that Joule drew the conclusion that the heat respectively evolved and absorbed in the first and third series was the equivalent of the work done respectively on and by the gas in these experiments, and thus afforded an independent proof of the convertibility of heat into mechanical effect, as well as a verification of the close approximation to the true value of his previous determination of the mechanical equivalent.

As regards Joule's position in the history of the discovery of the conversion of heat and work, the appreciation which he showed of the possibility of phenomena of air being complicated by latent heat, and his proof of its non-existence, are very important. It is certain that at this time Joule was unaware that Mayer had in 1842 brought forward the heat absorbed in the expansion of air as in itself affording a proof of the convertibility of heat into work. Had he done so he could not have conceived experiments better contrived to show the hypothetical character of the proof assumed by Mayer than those described in this paper, or have furnished those who, in the discussion which arose years afterwards, defended his position with more crushing arguments.

The conclusion of this paper has great interest as containing the expression of facts and opinions, the influence of which on Joule's subsequent work was considerable. Speaking of these experiments, he says that they "afford a new and, to my mind, powerful argument in favour of the dynamical theory of heat, which originated with Bacon, Newton, and Boyle, and has been, at a later period, so well supported by the experiments of Rumford, Davy, and Forbes. With regard to the detail of the theory, much uncertainty at present exists. The beautiful idea of Davy that the heat of elastic fluids depends partly on a motion of particles round their axes, has not, I think, hitherto received the attention it deserves. I believe most phenomena may be explained by adapting it to the great electro-chemical discovery of Faraday, by which we know that each atomic element is associated with the same absolute quantity of electricity. Let us suppose that these atmospheres of electricity, *endowed, to a certain extent, with the ordinary properties of matter*, revolve with great velocity round their respective atoms, and that the velocity of rotation determines what we call temperature. In an aeriform fluid we may suppose that the attraction of the atmospheres by their respective atoms, and that of the atoms towards one another, are inappreciable for all pressures to which the law of Boyle and Marotte applies, and that, consequently, the centrifugal force of the revolving atmospheres is the sole cause of the expansion on the removal of pressure. By this mode of reasoning the law of Boyle and Marotte receives an easy explanation without recourse to the improbable hypotheses of a repulsion, varying in a ratio different from that of the inverse square. The phenomena described in the present paper, as well as most of the facts of thermo-chemistry, agree with this theory; and in order to apply it to radiation, we have only to admit that the revolving atmospheres of electricity possess, in a greater or less degree, according to circumstances, the power of exerting isochronal undulations in the ether which is supposed to pervade space."

"The principles I have adopted lead to a theory of the steam engine very different from the one generally received, but at the same time much more in accordance with facts. It is the opinion of many philosophers that the mechanical power of the steam engine arises simply from the passage of heat from a hot to a cold body, no heat being necessarily lost during the transfer. This view has been adopted by Mr. E. Clapeyron in a very able theoretical paper, of which there is a translation in the third part of Taylor's "Scientific Memoirs." This philosopher agrees with Mr. Carnot in referring the power to *vis viva*, developed by the caloric, contained by the vapour, in its passage from the temperature of the boiler to the temperature of the condenser. I conceive that this theory, however ingenious, is opposed to the recognised principles of philosophy, because it leads to the conclusion that the *vis viva* may be destroyed by an improper distribution of the apparatus. Thus Mr. Clapeyron draws the inference that 'the temperature of the fire being from 1000°(C) to 2000° (C) higher than the boiler, there is an enormous loss of *vis viva* in the passage of the heat from the furnace to the boiler.' Believing that the power of destroying things belongs to the Creator alone, I entirely coincide with Roget and Faraday* in the opinion that in any theory which, when carried out, demands the annihilation of force, is necessarily erroneous. The principles, however, which I have advanced in this paper are free from this difficulty. From them we may infer that the steam while expanding in the cylinder loses heat in quantity exactly proportional to the mechanical force which it communicates by means of the piston, and that on condensation of the steam the heat, thus converted into power, is not given back. Supposing no loss by radiation, etc., the theory here advanced demands that the heat given out in the condenser shall be less than that communicated to the

---

* It was perpetual motion not annihilation of force—a very different thing—to which these objected.—*See* Note A.

boiler from the furnace, in exact proportion to the equivalent of mechanical power developed."

"It would lengthen this paper to an undue extent were I now to introduce any direct proofs of these views, had I even leisure at present to make the experiments requisite for the purpose ; I shall, therefore, reserve the further discussion of this interesting subject for a future communication, which I shall hope to have the honour of presenting to the Royal Society at no distant period,"

The opinions expressed in the several paragraphs in this remarkable quotation require separate consideration, as possessing, each of them, importance in connection with Joule's subsequent work, and his priority in expressing views now accepted as correct.

In the first place, it should be noticed that the two last paragraphs contain the first expression Joule recorded of an opinion, or even suggestion, that the work done by the steam engine is at the expense of the heat received from the fire. In describing his discovery of the "Mechanical Value of Heat," Joule had referred to the proportion between the work done by these engines, as compared to the whole mechanical equivalent of the heat of the coal consumed, but he there says not a word as to the conversion of heat into work by the engine. It might or might not have been in his mind, for anything he says. It has already been pointed out, however, that the paper contains evidence that he had at that time considered electricity as the only means of converting heat into work. But whether the suggestion was in his mind at the time, or followed immediately after writing the paper, he, for more than a year, refrained from expressing it until he had convinced himself and was in a position to convince others by the results of his experiment on the condensation of air. He now advances it boldly, clearly, and fully, and yet in no way over-states his case from the present point of view.

The reference to, and discussion of, Carnot's theory, which is contained in the second paragraph, is also of great interest. It is the first reference by any of those who engaged in the development of the general theory of thermo-dynamics to Carnot's theory, which afterwards attracted so much attention. It shows that Joule was at this time aware of, and had to some extent studied, Carnot's papers. But, what is of far greater interest and importance, it shows, by the light subsequently thrown on the subject, the difficulty Joule was labouring under in his task of reconciling his discovery with the apparently conclusive but contrary testimony of the steam engine itself, arising from his own want of appreciation of the limits imposed by surrounding conditions to the possibilities of converting heat into mechanical power. It also brings out the difficulties arising from want of common language in which to express newly-discovered facts. Carnot had, in 1824, discovered and proved conclusively an occult, but fundamental, principle respecting the relation of surrounding circumstances to the possibilities of obtaining power by means of heat, a principle which, as expressed by him, was in strict accordance with the common experience with the steam engine. In expressing this principle, Carnot had to confine himself to the knowledge of his time and to the language in which this knowledge was then expressed, however imperfect he himself might perceive this knowledge to be. The hypothesis that heat or caloric was indestructible and uncreatable, on which the proof of Carnot is based, in itself, contained no error, but the deduction from it, then generally made, that the heat received by the engine from the fire must therefore be equal to the heat given out to the condenser was erroneous. In the proof given by Carnot of his theory this deduction enters needlessly, but in illustrating the application of his theory to the steam engine he has explicitly accepted it, though not without protest.

Joule, reading Carnot—twice interpreted—seizes on the error of Carnot's illustration as evidence of the existence of views amongst philosophers which are inconsistent with the discoveries which he has made, and as serving to point the importance of the applications of these discoveries. This is only natural, and there is not a word in what Joule says which is not now accepted as strictly true. But what is remarkable is that Joule should have entirely ignored the truth and force of Carnot's theorem, and failed to observe that his own discovery cleared up the only thing obscure in Carnot's, and rendered this as intelligible as it was sound. Had Joule done this—had he observed that the theorem of Carnot was based solely on the indestructibility of caloric in its then accepted sense, as including latent as well as free heat, and that according to his (Joule's) own explicit statements, mechanical power had been previously included in latent heat as in the case of steam, in which the latent heat, as then understood, included the power spent in overcoming the pressure of the atmosphere—he would have seen that the performance of "work" was essentially a transmission of caloric in the accepted use of the term ; and that so far from refuting the hypothesis as to the indestructibility and uncreatability of caloric he had fully established this hypothesis and given it a rational explanation. The term 'caloric' would not then have been discredited, and would still be used in place of 'energy,' to which place it had not only the right given by antiquity, but was also much better fitted. In saying that Joule's missing this opportunity was remarkable, it is not intended to imply that it was in any way discreditable to his scientific insight. Joule was, at that time, head and shoulders before any of his contemporaries in his appreciation of matters relating to the connection between work and heat; and it was not till three years afterwards that Sir William Thomson, then commencing his intimate friendship with Joule, recognizing the incontestable evidence afforded by Joule's experiments, and anxious, beyond measure, to reconcile the apparent disagreement in the two principles, Joule's and Carnot's, was yet for three more years baffled by their apparent inconsistency, which arose from what was really a confusion of terms. It was then Sir William applied Young's term 'Energy' to include everything resulting from or convertible into the half of *vis viva*. Joule had used the term 'force' in the same sense as 'energy,' and continues to use it subsequently, quoting in defence Leibnitz's definition : 'The force of a moving body is proportional to the square of its velocity, or to the height to which it would rise against gravity.'

The first paragraph quoted from Joule's paper of June, 1844, is of great interest as showing his early appreciation of the bearing of his discoveries on the then very indefinite "dynamical theories" of "*heat*" and of "*gases*," and as containing an account of his own highly philosophical and successful attempt to add definition to the previous ideas of the motions of the matter constituting gases expressed by succeeding philosophers, which ideas had taken intelligible but still indefinite shape in Davy's hypothesis.

But from a biographical point of view the paragraph

DYNAMICAL SIGNIFICANCE OF "WORK." 93

has a still higher interest in that it shows how and by what means the skin of familiarity was at last pricked, and Joule's curiosity excited as to the dynamical significance of "work."

His papers have shown how little Joule thus far owes to whatever knowledge he may have initially had other than that resulting from his own familiarity with the subjects and means of his research : how he has been guided entirely by the curiosity which his own observations and discoveries have excited in his mind ; and how successively having his attention turned on to the relations between the, to him, initially familiar mechanical power as measured by " work," and the several physical quantities, magnetism, electricity, heat, and chemical affinity, and as to the inter-relations between these physical quantities, he has succeeded in realising an intimate knowledge of these physical quantities, of the inter-relations between them, and of the several relations between them and mechanical power—a knowledge including all that was known at his time, but extending far beyond.

Now, for the first time, his attention is directed to the relations between mechanical effect, as measured by the product of motion multiplied by resistance and other measures of mechanical action applicable to express the mechanical effect represented by a moving body in terms of its weight and motion.

That such measures had been fully investigated, and were known to all mathematicians, must have been known to Joule, and he must have had some vague ideas of these measures ; but he had received no education in the only language in which they were expressed, and they had not hitherto excited his curiosity.

He proceeds in this case as he has proceeded in all the

previous, expressing his ideas in suggestive but somewhat vague language. But this procedure, which was the only course where the ideas were entirely new, as now applied to old ideas, for which definite expressions have been adopted, not only conveys a wrong impression as to his meaning, but lays him open to the criticism of ignorance of the first principles of mechanical philosophy.

The expression (used in the paragraph already quoted) "that the velocity" of the whirling atmospheres, by which he assumes the respective atoms of the gas surrounded, "determines what we call the temperature," suggests, but does not fully justify, such criticism. But he has given a previous description of his hypothesis of whirling atmosphere in an appendix added on February 20th, 1844, to his paper of January 4th, 1843, on "The Heat Developed by the Electrolysis of Water." And here he not only makes a similar statement, but goes on to say that the *momentum* of the atmospheres constitutes "caloric," and later on says "that the momentum of the revolving atmospheres of electricity in a pound of water at freezing is equal to a mechanical force able to raise a weight of about 400,000lb. to a height of one foot." If he used the term *momentum* in its accepted mechanical sense as implying the product of mass mutiplied by velocity, these passages certainly show that he was, at that time, in error as to the principles of mechanics. That he did use *momentum* in this sense is implied in his definition of temperature quoted above. It is confirmed also by the fact that a year later, at the conclusion of another paper, he goes somewhat out of his way to recur to this subject, pointing out definitely that the "*vis viva*" of the atmosphere will be proportional to the square of the velocity, and then follows with " We see then what an enormous quantity of *vis viva* exists in matter. A single pound of water must possess 508° of heat, or in other words it must possess a *vis viva* equal to that acquired by a weight of 415,036lb. after falling through a perpendicular height of one foot." Joule does not here, or anywhere else in his subsequent writings, make reference to having used *momentum* and *vis viva* in the same sense. In whatever sense he used it, however, it makes little difference. The fact that he used it in other than its full mechanical sense shows that he was unacquainted with mechanical philosophy when he framed his hypothesis. While it is equally clear that in considering the results of that theory he, within the space of a year, realized this philosophy in its full significance, and found his hypothesis in strict accordance.

Joule's hypothesis of whirling atmospheres itself deserves more than passing notice. It was the first definite hypothesis of the kinetic constitution of gas, and although Joule made no attempt at its full mathematical development, the same theory, with but slight modification, was revived five years later by Rankine, under the title of " *The Hypothesis of Molecular Vortices,*" and in its mathematical development led him to the complete mathematical theory of the 'Mechanical Action of Heat' as it exists at present. This hypothesis, however, though explaining all the principal kinetic properties of gases, eventually failed to account for the properties of diffusion and viscosity, then but little known, but upon which Graham was engaged in his life-long researches; while the hypothesis that the particles of gas are flying about in all directions, discussed by Daniel Bernouilly, revived by Herapath in 1847, and developed first by Joule, being found capable of explaining the results of Graham's researches, as well as all the properties of gases explained by Davy's hypothesis, the latter was subsequently abandoned after it had done excellent service.

In the first instance while still using "*momentum*" Joule was led by this hypothesis to the solution of one of the oldest fundamental questions in science by his discovery of the *absolute zero of temperature.* Since the pressure of gas in a closed vessel is proportional to the temperature, and as both of these quantities are by Joule's definition of Davy's hypothesis proportional to the heat in the gas, it followed as a simple deduction that the absolute zero of temperature corresponded with the temperature of zero of pressure as indicated on the gaseous thermometer. This value Joule gave in the first statement of his theory, made in the appendix referred to, as 480 degrees Fahrenheit below freezing point. And although, on account of a small error in the determination of the co-efficient of expansion of air, on the publication of Regnault's determination of this co-efficient in 1848, this value had to be corrected by Joule from 480 to 493° below freezing point, to Joule belongs the credit of having discovered the existence and accurately determined the position of the *absolute zero of temperature.*

Nor was this all Joule learned by means of his hypothesis. According to this the specific heat of a body is proportional to the number of atoms in combination divided by the atomic weight. It thus directed his attention to the laws connecting the specific heats of gases and solids, and the weights and composition of the molecules; leading him first to study and discuss all the work then done on this subject, from Haycraft to Regnault, and then to compare the results with his theory. Which comparison, showing a remarkable agreement, he read before the British Association, September, 1844.

FIRST USE OF THE PADDLE.

Having his attention thus directed to the determination of specific heats, he invented two new methods of making these determinations, which he described in the Manchester *Memoirs*, 1845 and 1847.

In the meantime Joule had made a fresh determination of the mechanical equivalent of heat by turning a paddle in a can of water, the result being "that for each degree of heat evolved by the friction of water a mechanical power equal to that which can raise a weight of 890lbs. to the height of one foot had been expended."

His description of this apparatus and these results is given in a letter to the *Philosophical Magazine*, August 6th, 1845. This letter is interesting in consequence of this method, which in spite of its having given him, in this his first attempt, the least accurate result of any of his methods, is the one by which he subsequently accomplished the determination with such extreme accuracy. The letter is also interesting because it contains distinct evidence as to his having now acquainted himself with the principles of mechanics, and this by means of his dynamical hypothesis of gases. He uses "mechanical power" instead of "force," which he has been using ever since his first statement of his equivalent in 1843. And he then goes on to discuss the result of his hypothesis, as showing the enormous quantities of heat that exist in matter at ordinary temperatures. After discussing the various values of his equivalents as derived 1st, from magneto-electrical experiments (823lb.); 2nd, from the cold produced by rarefaction of air (795lb.); 3rd, from experiments on the motion of water through narrow tubes (774lb.), and with the paddle 890lb., and obtaining as the mean of the three classes of experiments 817, he says:—

H

"Admitting the correctness of the equivalent I have named, it is obvious that the *vis viva*" (the first time Joule has used this expression in connection with his own experiments) "of the particles of a pound of water at (say) 51°, is equal to the *vis viva* in a pound of water at 50°, plus the *vis viva* that would be acquired by a weight of 817lb. after falling through the perpendicular height of one foot." Even now he does not use *vis viva* according to its mathematical definition, but as half the quantity.

"Assuming," he continues, "that the expansion of elastic fluids is owing to the centrifugal force of revolving atmospheres of electricity we can easily estimate the absolute quantities of heat in matter. For in an elastic fluid the pressures will be proportional to the square of the velocity of the revolving atmospheres, and the *vis viva* of the atmospheres will also be proportional to the square of the velocity. Consequently the pressure will be proportional to the *vis viva*. Now the ratio of the pressure of the elastic fluids at temperatures 32° and 33° is 480÷481, consequently the zero of temperature must be at 480 below the freezing point of water. We see then what an enormous quantity of *vis viva* exists in matter. A single pound of water at 60° must possess 480+28=508° of heat, or, in other words, it must possess a *vis viva* equal to that acquired by a weight of 415,036lbs. after falling one foot. The velocity with which the atmosphere of electricity must revolve in order to present this enormous amount of *vis viva* must of course be prodigious, and equal probably to the velocity of light in the planetary space, or to that of an electric discharge, as determined by the experiments of Wheatstone."

This account of the application of his theory, which is expressed in strict accordance with dynamical principles, is remarkable as containing the first definite dynamical deductions ever made from gaseous hypothesis. It is also the last mention Joule makes of this hypothesis, except the mere mention, which he makes in 1848, in stating that he adopts the hypothesis of Herapath, published in 1847, as being simpler.

Although Joule's earlier researches, those made before August, 1843, which led him to the discovery of the principal relations on which the law of conservation of energy is based, and during which he acquired his unique facility in quantitative measurement, must ever remain unrivalled in interest and importance by any of his subsequent work, it is interesting to notice that his capacity for experimental work has so far been steadily increasing.

In the interval from the summer of 1843 to that of 1845, besides the researches already mentioned, he was engaged on two other very considerable researches having a direct bearing on the main line of his work.

After Joule's discovery of the mechanical equivalent of heat in 1843 the electro-magnetic investigation which he was continuing with Dr. Scoresby was invested with a new interest, and was continued with a view to comprising the economic capabilities of electro-magnetism, steam, and horses, as sources of power, and the comparison of the "work" done with the heat equivalents of the zinc, coal, and food, severally consumed by these sources of power. In 1845, during the research, Joule spends some time visiting Dr. Scoresby, at Bradford. The results of this research were published in the *Philosophical Magazine* in 1846, under the title :—

"*Experiments and Observations on the Mechanical Powers of Electro-magnetism, Steam, and Horses. By the*

Rev. William Scoresby, D.D., F.R.S.S.L. and E., Corr. Mem. Inst. Fr., &c., and James P. Joule, Secretary of the Literary and Philosophical Society of Manchester, Mem. Chem. Soc. &c."

To the reprint of this paper in 1885, Joule adds the following interesting note :—

"On the occasion of the meeting of the British Association for the Advancement of Science, at Manchester, in the year 1842, I had the happiness of forming the acquaintance of Dr. Scoresby, eminent for qualities seldom united in one man. At once an experienced seaman, a successful geographical discoverer, a hard working and eloquent clergyman, he was also a zealous student of nature and a scientific investigator. Dr. Scoresby became greatly interested in the view I was at that time beginning to take of the relation between heat and other forms of force, and in response to my express wish to work with a powerful arrangement of magnets, he kindly invited me to Bradford, of which town he was at the time Vicar, in order to pursue an inquiry along with him. The duties of the parish were, however, so onerous and pressing, that the production of our paper devolved almost entirely on myself, so that it was not without great objection on his part that Dr. Scoresby allowed his name to appear with mine. Inasmuch, however, as the facilities for the experiments were afforded by him, as well as the great magnetic battery, I felt that I could not, in justice, allow it appear other than as a joint paper."

The final results of this investigation, which are of great interest, may be shortly summarised as follows :—The duty of an electro-magnetic engine per grain of zinc consumed in a Daniell's battery, is 80lb. raised one foot high, about half the theoretical maximum duty. The duty of a

Cornish engine at that time, per grain of coal, is 143lb. raised one foot high, or one-tenth the *vis viva* due to the combustion of coal.

The duty of a horse per grain of food is 143lb. raised one foot high, or one quarter of the *vis viva* resulting from the combustion of the food.

During the interval between the summer of 1844 and the end of 1845, Joule is also deeply engaged in repeating and extending his research made in 1842 on the heat evolved during electrolysis.

The results of this extended research he communicates in an essay to the French Academy of Sciences, under the motto :—

"Actioni contraria semper et aequalis est reactio."—*Newton*. in competition for the prize offered for the best essay on " The Heat of Chemical Combinations." The merit of this essay is unquestionable, but Joule seems to have been unaware of the regulations to be observed by the competitors for the prize, and to have been too late. The result being that his essay was not published till 1852, when he sent it to the *Philosophical Magazine*.

In 1846, besides working at the perfection of his apparatus and experiments for obtaining with greater accuracy the mechanical equivalent of heat, Joule is also engaged on two very heavy experimental researches.

Having completed the research on magnetism, which he had undertaken with Dr. Scoresby, he takes up again " The Effect of Magnetism upon the Dimensions of Iron and Steel." This he had begun in 1841, when Mr. F. D. Arstall suggested to him a new form of electro-magnetic engine based on this, then supposed, effect. The first results were published in Joule's first public lecture (already mentioned).

In this lecture he "made it evident that an increase of length in a bar of iron was produced by magnetizing it, and therein also states his reasons for believing that, whilst the bar was increased in length by the magnetic influence, it experienced a contraction at right angles to the magnetic axis, so as to prevent any change taking place in the bulk of the bar. In the meantime this inquiry has been taken up by De la Rive, Matteucci, Wertheim, Wartmann, Marrian, Beatson, and others, whose ingenious experiments have invested the subject with additional interest." Joule now made a complete investigation, which he published in the *Philosophical Magazine*.

The other investigation which Joule was engaged upon at this time was on "Atomic Volume and Specific Gravity," which he undertook in conjunction with Sir Lyon Playfair. Of this investigation, Joule says in a note to the reprint in 1885:—

"My work with Dr., now the Right Hon. Sir Lyon Playfair, K.C.B., was commenced at about the time when he occupied the post of Chemist to the Royal Manchester Institution. It is only just to observe that the important theoretical results arrived at with regard to atomic volumes, are almost entirely due to him, while I took the principal part in the experiments on the expansion of salts, the maximum density of water, &c. My own individual work was done at Oak Field House, Upper Chorlton Road, Manchester."—*Note*, 1885, J. P. J.

The amount of experimental work done in this investigation, which was completed in 1847, was very great; the results being communicated in no less than seventeen papers to the Chemical Society.

CHAPTER VII.

THE YEAR 1847.—*Lecture at St. Ann's Church Reading Room.—Conservation of "Force."—Fresh Determination of Equivalent.—Verification of Laplace's Theory of Sound. —Joule's Paper accepted by the Institute of France.— Meeting of British Association at Oxford.—First Public Recognition of Joule's Discoveries.—Joule's Account.— Sir William Thomson's Account.—Marriage.—Shooting Stars.—The Adoption of Herapath's Hypothesis.—Determination of Velocities of Molecules of Gases and Theoretical Specific Heats.*

1847 proved a most important year in the life of Joule, although at its commencement, whatever reason he may have had to expect that it was to be the year of his marriage, he had no reason to foresee the fortunate influence on his scientific career which the events of this year were to exercise. By the end of 1846 he had completed nearly all the investigations he had been engaged upon. He had made himself master of the significance of the principle we now call energy in all its physical and mechanical modes, and he had received his first scientific honour—the highest this society could then award him—in his election to the office of secretary. Thus, whatever it might prove in his family relations, 1847 must have opened to him as likely to be a quiet year in his scientific work.

On the 28th April, 1847, Joule gave a popular lecture in

Manchester, at St. Ann's Church Reading Room, "On Matter, Living Force, and Heat," which lecture was published in full in the *Manchester Courier* newspaper, May 5 and 12, 1847, and has already been quoted in Chapter I.

In this lecture, Joule gave the first full and clear exposition of the universal conservation of that principle now called energy. He does not use the term energy; nor, having abandoned the term caloric, has he any name for the principle; but he, nevertheless, in very forcible language, shews that he recognises the quantitative measures of the principle in all the "mechanical, chemical or vital" sources of heat as yet known; and further recognises that all the phenomena of the universe consist of the continual conversion of the principle from one of its modes into another without loss or increase. It is very evident throughout this lecture that Joule had then no clear conception of the agency in the conversion of energy, of what is now expressed as concentration of energy, much less had he any conception that the conversions of the energy, in which all the phenomena of the universe consist, are effected largely at the expense of this agent in accordance with Carnot's theorem. But this ignorance—the ignorance of his time—though it rendered the expression of his own clear conception of the law of the conservation of energy difficult, does not in any way invalidate the completeness and truth of this exposition of the grandest generalization in the whole of physical science.

This exposition, besides being the first ever given, was the only one Joule ever gave, except such as may be gathered from the general tenour of his writings. That he attached great importance to the lecture appears from the following note by Mr. Benjamin Joule.—"James was very anxious that this lecture should be published in its entirety as soon as possible. One paper refused to give even a notice of it. After some discussion the *Manchester Guardian* would, as a favour, print extracts to be selected by themselves. This of course would not satisfy my brother. I returned to the *Manchester Courier*, and, after a long debate, they promised to insert the whole as a special favour to myself."

Although thus fortunately preserved, the lecture was lost sight of till 1884, so that Joule's generalization was not before the scientific world. Then, after about five years, as a consequence of the recognition of Joule's discoveries, the generalization was borne in upon the minds of those who took up his work; and, when expressed by Sir William Thomson as the law of conservation of energy, it was accepted as one of the fundamental laws of the universe.

It thus becomes clear that this lecture marks an epoch in the history of science, and that, besides the debt of gratitude owing to Joule for making the discoveries which rendered this generalization possible, there is owing to him the further debt for having himself propounded the generalization.

In this lecture also Joule mentions, for the first time, by way of illustration, his now accepted explanation of Shooting Stars :—That they are meteorites rendered hot by the friction they meet on encountering the atmosphere. This explanation, to which Joule was directly led by his experiments on the heat developed by the friction of fluids, forms the subject of a future communication on Shooting Stars.—*Philosophical Magazine*, 1848.

In the meantime, Joule had been making fresh determinations of the mechanical equivalent of heat by experiments on fluid friction with his now nearly perfected apparatus, using sperm-oil as well as water, and had obtained 781·8 as the equivalent resulting from nine experiments with water, and 782·1 as resulting from nine experiments with sperm-oil. He communicated these results in June, 1847, to the chemical section of the British Association.

In a letter to the *Philosophical Magazine*, dated Oak Field, near Manchester, July 17th, 1847, Joule says—"The equivalent of a degree of heat per lb. of water, determined by careful experiments made since those brought before the British Association at Oxford is 775lb. through a foot!"

This letter is primarily "On the Theoretical Velocity of Sound." This, according to Laplace, depends on the ratio of the specific heat of air at constant pressure and that at constant volume.

The values of these specific heats, already determined, at that time were known to be inexact, particularly that at constant volume. Joule now sees that he can, by means of the mechanical equivalent of heat, determine the value of specific heat at constant volume relatively to that at constant pressure with a close approximation. He uses the new equivalent 775 in making this calculation, and obtains the ratio 1·36. This is not far from the true value 1·408; the difference being solely attributable to the error of the specific heat of air at constant pressure. Using this ratio in the formula of Laplace, Joule found that it brings up the velocity of sound "from Newton's estimate of 943 to 1095, which is as near 1130, the actual velocity at 32°, as could be expected from the nature of the experiments on the specific heat, and fully confirms the theory of Laplace." This determination was in a few years to

THE FIRST RECOGNITION.

have an importance besides that which naturally belonged to it.

On August 23rd, 1847, there appeared in the "Comptes Rendus" a memoir, entitled, "*Experience sur l'Identité entre le Calorique et la Force Mécanique. Détermination de l'équivalent par la Chaleur dégagée pendant la friction du Mercure. Par M. J. P. Joule.*"

Joule adds a note, 1881 :—"The commissioners were Biot, Pouillet, and Regnault. I had the honour to present the iron vessel, with its revolving paddle-wheel, to the last-named eminent physicist."

The Institute of France was therefore the first of any national academy or institute to recognise the importance of Joule's discoveries.

On the 23rd of June, 1847, the British Association assembled in Oxford, and the meeting will ever be memorable as the occasion on which Joule's physical discoveries received their first recognition; just as the next meeting in Oxford, 1860 is memorable as the occasion at which the tide of public opinion was turned in favour of Darwin's theory of evolution, by Huxley's defence against the attacks of Wilberforce.

Writing in 1885 Joule thus describes what occurred : —

"It was in the year 1843 that I read a paper 'On the Calorific Effects of Magneto-Electricity and the Mechanical Value of Heat' to the Chemical Section of the British Association assembled at Cork. With the exception of some eminent men, among whom I recollect with pride, Dr. Apjohn, the president of the section, the Earl of Rosse, Dr. Eaton Hodgkinson, and others, the subject did not excite much general attention, so that when I brought it forward again at the meeting in 1847 the chairman suggested that, as the business of the Section pressed, I should not read my paper but confine myself to a short verbal description of my experiments. This I endeavoured to do, and a discussion not being invited the communication would have passed without comment if a young man had not risen in the section, and by his intelligent observations created a lively interest in the new theory. The young man was William Thomson, who had two years previously passed the University of Cambridge with the highest honour, and is now probably the foremost scientific authority of the age."

Writing to Mr. J. T. Bottomley in 1882 (*Nature*, Vol. XXVI., p. 618), Sir William Thomson also writes of this meeting :—

"I made Joule's acquaintance at the Oxford Meeting, and it quickly ripened into a life-long frendship. I heard his paper read at the sections, and felt strongly impelled to rise and say that it must be wrong, because the true mechanical value of heat given, suppose to warm water, must, for small differences of temperature, be proportional to the square of its quantity. I knew from Carnot's law that this must be true (and it *is* true; only now I call it 'motivity,' in order not to clash with Joule's 'Mechanical Value'). But as I listened on and on, I saw (that though Carnot had vitally important truth not to be abandoned) Joule had certainly a great truth and a great discovery, and a most important measurement to bring forward. So instead of rising with my objection to the meeting, I waited till it was over and said my say to Joule himself at the end of the meeting. This made my first introduction to him. After that I had a long talk over the whole matter at one of the conversaziones of the Association, and we became friends from thence forward. However, he did not tell me he was to be married in a week or so, but about a fortnight later, I was walking down from Chamounix to commence the tour of Mont Blanc, and whom should I meet walking up but Joule, with a long thermometer in his hand, and a carriage with a lady in it not far off. He told me that he had been married since we parted at Oxford! and he was going to try for elevation of temperature in waterfalls. We trysted to meet a few days later at Martigny, and look at the Cascade de Sallanches, to see if it might answer. We found it too much broken into spray. His young wife, as long as she lived, took complete interest in his scientific work, and both she and he showed me the greatest kindness during my visits to them in Manchester, for our experiments on the thermal effects of fluid in motion, which we commenced a few years later.

"Joule's paper at the Oxford meeting made a great sensation. Faraday was there, and was much struck with it, but did not enter fully into the new views. It was many years after that, before any of the scientific chiefs began to give their adhesion. It was not long after when Stokes told me he was inclined to be a Joulite.

"Miller or Graham, or both, were for many years quite incredulous as to Joule's results, because they all depended on fractions of a degree of temperature—sometimes very small fractions. His boldness in making such large conclusions from such very small observational effects, is almost as noteworthy and admirable as his skill in extorting accuracy from them. I remember distinctly at the Royal Society, I think it was either Graham or Miller saying simply he did not believe Joule because he had nothing but hundredths of a degree to prove his case by."

In this letter it may be assumed that Sir William Thomson does not include himself as amongst the then scientific chiefs (although it is hard to believe he was not) who withheld their adhesion for many years, but he is careful not to say to what extent he at once followed Joule. This, however, he has clearly shown by his writings. He accepted the results of Joule's experiments on the heat produced by fluid friction as proved; but he did not for two years follow Joule in the general significance which Joule attached to them. The principle of Carnot stood in his way: he could not reconcile this with the general view of the reciprocal convertibility of heat and work, and until he had done so he could not pass it by. Sir William Thomson had only as yet realised one of Joule's results, and during the two years of doubt he was working backwards along the path which Joule had trodden until he had mastered the whole of Joule's work, and then he not only realised the revelation which Joule had made but received his own inspiration of the general physical significance of Carnot's theory and the underlying principle of dissipation of energy.

From these quotations it is clear that both Sir William Thomson and Joule could look back to the occasion of their meeting as an extremely happy one.

To Joule it was one of the happiest circumstances of his life as well as having most important influences on the recognition of his work.

In June, 1847, James Prescot Joule was married to Amelia, daughter of Mr. John Grimes, Comptroller of Customs, Liverpool, at St. Peter's Church, Rock Ferry, Higher Bebington, by the Rev. Thomas Fisher Redhed, incumbent, Mr. Dawson giving away the bride. The bride and bridegroom started for Switzerland, taking London and Paris in their way. They returned to Oak Field, 23rd September, to reside till 9th April, 1849, when they entered their house in Acton Square, Salford.

Joule continued his experiments for the mechanical equivalent of heat in his laboratory at Oak Field.

He was also pursuing the theoretical extension of the dynamical theory of heat resulting from his discoveries.

Early in 1848 he communicates to the *Philosophical Magazine* the paper "On Shooting Stars," already referred to. In this paper he makes a definite estimate of the resistance encountered by a meteorite of the size of a six inch cube, moving at the rate of eighteen miles a second, through an atmosphere of one hundredth the density of that at the earth's surface, and finds this to be 510,600lb. He then determines the heat equivalent of this resistance, if the stone traverses twenty miles, equal to raising 6,969,980lb. of water 1° Fahrenheit. He then continues :—

"Of course, the larger portion of this heat will be given to the displaced air, every particle of which will sustain the shock, while only the surface of the stone will be in violent collision with the atmosphere. Hence the stone may be considered as placed in a blast of intensely heated air, the heat being communicated from the centre to the surface by conduction. Only a small portion of heat will be received by the stone, but if we estimate it at only one hundredth, it will still be equal to 1° Fahrenheit per 69,679lb. of water, a quantity quite equal to the melting and dissipation of any materials of which it may be composed.

"It appears to me that the varied phenomena of meteoric stones and shooting stars may all be explained in the above manner, and that the different velocities of the aerolites, varying from four to forty miles per second, according to the direction of their motion with respect to the earth, along with their various sizes, will suffice to show why some of these bodies are destroyed the instant they arrive at our atmosphere, and why others arrive at the earth's surface with diminished velocity.

"I cannot but be filled with admiration and gratitude for the wonderful provision thus made by the Author of Nature for the protection of His creatures. Were it not for the atmosphere which covers us with a shield, impenetrable in proportion to the violence which it is called upon to resist, we should be continually exposed to a bombardment of the most fatal and irresistible character. To say nothing of the larger stones, no ordinary buildings could afford shelter from very small particles striking at the velocity of eighteen miles per second. Even dust flying at such a velocity would kill any animal exposed to it."

In the summer of 1848 Joule had completed a very bold and successful attempt to submit the dynamical theory of gases to the test of experiment, and communicated the results to this Society. In this attempt he accepts the theory of Herapath as simpler than his own. He gets over the mathematical difficulties by an assumption which shows his insight into the subject, as it is exactly that to which the mathematical solution of the problem would have led him; and from this assumption he calculates exactly what would be the velocity of the particles of hydrogen, supposed all to move with the same velocity, in order to give the, experimentally ascertained, pressure at a certain density and temperature of the gas. The result he obtains is, for a temperature of 32° Fahrenheit, a velocity of 6055· feet per second, while the most accurate determination that has yet been made gives it as 6049·.

VELOCITY OF GASEOUS MOLECULES. 113

From this he then correctly deduces the corresponding velocities for the molecules of oxygen, nitrogen, carbonic acid, and steam as being inversely proportioned to the square roots of these specific gravities.

He then estimates, by means of his discoveries of the dynamical equivalent of heat and absolute zero of temperature, what would be the specific heats of the several gases as resulting from the *vis viva* of these molecular velocities, supposing, according to Herapath's hypothesis, that the molecules are hard. These theoretical specific heats he compares with the experimental results of Delaroche and Berard, which he finds are invariably much higher and not in simple proportion. He has thus found that Herapath's hypothesis of hard particles is not capable of explaining the then experimental results, but as he expects the experiments undertaken by M. V. Regnault for the French Government will shortly be published, he says, "Till then it will be better to delay any further modifications of the dynamical theory by which its deductions may be made to correspond more closely with the results of experiment." Considering that these results were obtained only as a consequence of Joule's discoveries of the mechanical equivalent of heat, and the absolute zero of temperature, that it was more than four years before any further work was done on this subject, and that not only the mean velocities, but also the specific heats resulting from the linear motion of the molecules are in accordance with the now accepted theory, which theory is Herapath's, with an addition introduced, first by Clausius, solely to meet the deficiency of the specific heats discovered by Joule to belong to Herapath's hypothesis, it appears that this memoir establishes Joule's position as the founder of the modern quantitative dynamical

I

theory of gases as well as the quantitative dynamical theory of heat.

At the British Association in 1848, Joule communicated the result of further experiments made for the equivalent, with a slight alteration in the apparatus calculated to give greater exactness to the results. The result arrived at being 771, "which is," he says, "believed to be within one two hundredth part of the truth." This it is found to be.

Joule published nothing further in 1848 or 1849, but during this interval he was occupied in making his final investigation, with his now perfected apparatus, on the mechanical equivalent of heat.

In the meantime, events were occurring in other places, which, as being directly connected with the recognition of the truth and importance of Joule's work, require notice in this memoir.

CHAPTER VIII.

JOULE'S VIEWS ACCEPTED BY THOMSON, RANKINE, AND
CLAUSIUS.—*Effect of Publication of Regnault's Researches.
—Thomson's First Paper on Mechanical Effect by Thermal
Agency.—Maintain's Inconvertibility of Heat.—Note on
Joule's Views.—Conversion of Heat into Work Denied.—
Work into Heat Accepted. — Second Paper. — Greater
Deference. — Accepts Joule's Difficulties as to Carnot's
Axiom.—Thomson's Courage in Expressing his Doubts.—
Discovery of Dissipation of Energy.—Rankine's Hypothetical Theory of Heat.—Acknowledges Joule's Hypothesis.—
Hypothetical Foundation Obscures General Laws.—Accepts
Joule's Views.—Criticises Joule's Experiments.—Apology
and Acceptance of Joule's Equivalent.—Joule suggests
the form of Carnot's function.—Clausius Theory—Based
on Joule's and Carnot's discoveries—Contains Hypotheses.—Thomson's Third Paper— General Foundation
without Hypotheses—Enthusiastic Acceptance of All
Joule's Views.—Joule's Final Determination of the
Mechanical Equivalent.—Historic Sketch.—Cordial Recognition of the Views and Work of his Predecessors
and Contemporaries.*

Besides Joule's exposition of the law of conservation of that principle now called energy, and the meeting of Joule with Sir William Thomson, at Oxford, another event took place in 1847, which alone would serve to mark the year as important in physical science.

On April 26, "*Mémoires de l' Académie des Sciences, ae L'Institut de France, Tome* XXI." was published, and contained "*Relation des Expériences Enterprises par Ordre de Monsieur le Ministre des Travaux Publiques et sur la proposition de la Commission Centrale des Machines à Vapeur, Pour Déterminer les principales Lois et les Donn'ees Numériques qui entrent Dans le Calcul des Machines à Vapeur par M. V. Regnault.*"

The publication of the first ten Memoirs of Regnault's now classical researches, containing full and accurate experimental determinations of the numerical relations between volume, density, pressure, and temperature of air, and particularly the relation between the temperature, pressure, and latent heat of steam, unquestionably took an important, if not a first, place in the sudden interest which arose in the theory of heat engines.

Sir William Thomson, at the time his attention was caught by Joule's discovery of the heat produced by the friction of water, had already published some 26 papers on theoretical physics, notwithstanding that he had only taken his degree two years before. These papers were largely devoted to the theory of the conduction of heat; but although at the time he was not only acquainted with Carnot's theory, but firmly convinced of its truth, he had not touched in any of his papers on the subject of the production of mechanical effect by thermal means, nor does he touch upon this in the first ten papers which he published during the succeeding eight months; but in June, 1848, he read a paper at the *Cambridge Philosophical Society*, "On an Absolute Thermometric Scale founded on Carnot's Theory of the Motive Power of Heat, and calculated from Regnault's observations."

This was not, as might naturally be supposed, inspired by anything he had learned from Joule, but was obviously the immediate result of the publication of Regnault's researches, as affording data which enabled him to make the numerical comparisons of the degrees centigrade with those of the absolute scale which had been previously suggested to him by Carnot's theorem.

The paper itself is altogether pre-Joule, and serves well to illustrate the state of assured error into which the casuistry of the schools relating to caloric reduced the minds of their most powerful and philosophical thinkers.

It appears from a note, however, that his interview with Joule had produced an effect which had not been altogether obliterated by the intense occupation of the mean time.

In his description of the foundation of Carnot's theory, Sir William Thomson writes :—

"In the present state of science no operation is known by which heat can be absorbed, without either elevating the temperature of matter, or becoming latent and producing some alteration in the physical condition of the body into which it is absorbed; and the conversion of heat (or *caloric*) into mechanical effect is probably impossible,* certainly undiscovered. In actual engines for obtaining mechanical effect through the agency of heat we must consequently look for the source of power, not in any absorption or conversion, but merely in a transmission of heat."

The note to the asterisk over "impossible" is :—

"This opinion seems to be nearly universally held by those who have written on the subject. A contrary opinion, however, has been advocated by Mr. Joule, of Manchester; some very remarkable discoveries which he has made with reference to the generation of heat by the friction of fluids in motion, and some known experiments with magneto-electric machines, seeming to indicate an actual conversion of mechanical effect into caloric. No experiment, however, is adduced in which the converse operation is exhibited; but it must be confessed that as yet much is involved in mystery with reference to these fundamental questions of natural philosophy."

Clearly in the interval which elapsed between the meeting at Oxford and the writing of the above, Sir William Thomson had been otherwise engaged than in mastering Joule's researches.

This note of Sir William Thomson's contains the first comments upon Joule's work published by any physicist then, or afterwards, eminent. Coming, as it did, after five years of absolute silence, and standing, as it does, alone for another six months, it has great interest in Joule's biography.

In itself it certainly contains but little in the way of recognition of the results of Joule's 10 years' labour. It contains the mistaken denial that Joule had proved the conversion of heat into work; showing that Joule had been too hasty in supposing that the results which were obvious to him as a simple logical deduction from his experiment on air and the, previously proved, experimental relations between the pressures, volumes, and temperature would be equally obvious to his readers. It also contains the still more mistaken denial that any experiment had proved the conversion of caloric (then used as including latent heat with free heat) into mechanical effect, showing that the writer had overlooked the direct proof of this contained in Joule's magneto-electrical experiments—a proof, the beauty of which the same writer, some two years later, not only points out but dwells upon; while the only acceptance which this note contains is the discovery, and this in its narrowest sense, of the conversion of the mechanical effect into heat.

Notwithstanding the denials, however, the admission by such a physicist, however small in itself, was an indication of great things. It was the thin end of the wedge, or the first drop that indicated that the tide of highest scientific thought had overtopped the obstructing bank of preconceived opinions, which it forthwith began to sweep away.

Sir William Thomson's next paper, "An account of Carnot's theory of the Motive Power of Heat; with its Numerical Results derived from Regnault's Experiments on Steam," was read before the Royal Society of Edinburgh, January 2, 1849. This paper shows that the writer has, since the last paper, been studying and considering Joule's published researches, to which he now pays considerable deference. He still continues to use the assumption that heat is not converted into work as the foundation of Carnot's theorem, but precedes this assumption by the following qualification in the text of his paper: "The extremely important discoveries recently made by Mr. Joule, of Manchester, that heat is evolved in every part of a closed electric conductor moving in the neighbourhood of a magnet; and that heat is *generated* by the friction of fluids in motion, seems to overturn the opinion generally held that heat cannot be *generated* but only produced from a source, where it has previously existed either in a sensitive or latent condition."

" In the present state of science, however, no operation is known by which heat can be absorbed into a body without either elevating its temperature, or becoming latent and producing some alteration in its physical condition;"

(this ignores the fact that the latent heat of the battery was absorbed in Joule's electro-magnetic engine, without producing any change in the condition of the substance of the engine) "and the fundamental axiom adopted by Carnot may be considered as still the most probable basis for an investigation of the motive power of heat; although this, and with it every other branch of the theory of heat, may ultimately require to be reconstructed upon another foundation, when our experimental data are more complete. On this understanding and to avoid a repetition of doubts, I shall refer to Carnot's fundamental principle in all that follows as if its truth were thoroughly established."

Besides this general qualification, the paper contains two long notes in which the writer discusses and argues some of Joule's results. In the first of these, Joule's experiments, on the effect of magneto-electricity in producing heat, are shortly described, and subjected to the criticism that Joule had not explicitly discussed any possible change of heat in the inducing magnets, the writer coming to the conclusion that in all probability this was not much. This criticism is a mistake, as there were no means for the conduction of heat or electricity between the two systems except radiation, and that Joule had discussed. The second note has reference to the difficulty which arises in explaining what becomes of the *vis viva* that would have been produced in an engine when the temperature of the heat is similarly lowered by simply causing it to pass through the solid plate of a boiler from the fire to the water. This was the very point that Joule took hold of in the passage quoted, page 88, from Joule's paper "On the Changes of Temperature produced by the Rarefaction and Condensation of Air;" and now Sir William Thomson is struck by it, his attention apparently having been called to it by Joule's remark, to which he refers in the same note. The note is—"When thermal agency" is thus spent in conducting heat through a solid, what becomes of the mechanical effect which it might produce? Nothing can be lost in the operations of nature— "no energy can be destroyed!" (The sense is identical with Joule's—" Believing that the power of destroying belongs to the Creator alone, I entirely coincide with Roget and Faraday in the opinion that any theory which, when carried out, demands the annihilation of force, is necessarily erroneous." "The principles, however," Joule then continues, "which I have advanced are free from this difficulty.") Sir William Thomson then continues :—" What effect, then, is produced in place of the mechanical effect which is lost ? A perfect theory of heat demands an answer to this question ; yet no answer can be given in the present state of science. A few years ago a similar confession must have been made with reference to the mechanical effect lost in a fluid set in motion in the interior of a rigid closed vessel and allowed to come to rest by its own internal friction ; but in this case the foundation of the solution of the difficulty has been actually found in Mr. Joule's discovery of the generation of heat, by the internal friction of a fluid in motion. Encouraged by the example, we may hope that the very perplexing question in the theory of heat, by which we are at present arrested, will before long be cleared up."

"It might appear that the difficulty would be entirely avoided by abandoning Carnot's fundamental axiom ; a view which is strongly urged by Mr. Joule (at the conclusion of his paper 'On the Changes of Temperature produced by the Rarefaction and Condensation of Air'). If we do so however, we meet with innumerable other difficulties—insuperable without further experimental investigation and an entire reconstruction of the theory of heat from its foundation. It is indeed to experiment we must look, either for a verification of Carnot's axiom and an explanation of the difficulty we have been considering, or for an entirely new basis of the theory of heat."

This note contains the last doubt or difficulty ever expressed by Sir William Thomson as to the full acceptance of all Joule's views, and, not only so, these quotations from Sir William Thomson's papers on Carnot's theory, published in 1848 and 1849, contain the only expressions of doubt ever published as to the sufficiency and conclusiveness of Joule's demonstration of the reciprocal convertibility of heat and mechanical effect, or as to the conclusions which Joule himself had drawn from them as to the mechanical character of heat. The propriety of introducing them in this memoir thus becomes evident, and it is also evident that any apology to the author of these doubts is not only unnecessary but would be out of place.

The expression of these doubts, before any other physicist had made any admission as to the importance or taken any public notice whatever of Joule's work, shows that Sir William Thomson, besides having the courage beyond all others to publicly announce and acknowledge the importance of discoveries, the truth of which, however adverse to the views he then held, he could not deny, alone had sufficient confidence in his own judgment, knowledge, and scientific acumen to maintain the truth of what his judgment convinced him was true in the previous theory, and while admitting the force of the new theory, to withhold his full acceptance, until by clearly seeing wherein the confusing error lay, he was able to reconcile what was true in the old with the new theory.

The last note shows how strong was the temptation upon Sir William Thomson to dismiss Carnot's theory as erroneous in the same way as Joule had dismissed the hypothesis of caloric without seeing that he had really established it on a rational foundation. To the fact that Thomson resisted this temptation we undoubtedly owe the maintenance of Carnot's theory as constituting, together with Joule's law, the only perfectly general foundation for the science of thermodynamics; and perhaps, more important still, it was as the solution to his doubts that Sir William Thomson was led to perceive the general tendency of the energy in the universe to "dissipation," until all action ceases, and a uniform temperature prevails.

This grand generalisation, expressed as the "universal tendency in nature to the dissipation of energy," though distinct from the law of the universal conservation of energy, which Joule had exposed, was a necessary complement to reconcile the former with the observed phenomena of nature. In his exposition of his law Joule had said, "Nothing is lost," and, in the time of his doubt, Thomson had repeated, " Nothing can be lost in the operations of nature—no energy is lost; " experience all the same showed that something was lost and as the solution of his doubts, Thomson finds something is lost,—not energy, but concentration of energy.

In December, 1849, there was presented to the Royal Society of Edinburgh, a paper by William John Macquorn Rankine, on "The Mechanical Action of Heat, especially in the case of Gases and Vapours." As Rankine was then only known as a promising young engineer, 29 years of age, this paper, containing as it did the complete mathematical development of the modern dynamical theory of heat in almost all of its important applications, came as a surprise.

Rankine founded his theory on the "Hypothesis of Molecular Vortices"—"*that each atom of matter consists of a nucleus or central point enveloped by an elastic atmosphere which is retained in its position by attractive forces, and that the elasticity due to heat arises from the centrifugal force of those atmospheres revolving or oscillating about their nuclei or central points.*"

"According to this hypothesis," Rankine proceeds, "quantity of heat is the *vis viva* of the molecular revolutions or oscillations."

"Ideas resembling this have been entertained by many natural philosophers from a very remote period; but, so far as I know, Sir Humphrey Davy was the first to state the hypothesis I have described in an intelligible form. It appears since then to have attracted little attention till Mr. Joule, in one of his papers on the production of heat by friction, published in the *London and Edinburgh Philosophical Magazine* for May, 1845, stated it in more distinct terms than Sir Humphrey Davy had done. I am not aware, however, that anyone has hitherto applied mathematical analysis to its development."

It thus appears that Rankine acknowledges Joule's priority in adding the definition of this hypothesis. That he does not give Joule credit for the important developments he had made in determining the absolute zero of temperature and the law of specific heats, must be attributed to the fact that they were not included in the paper to which Rankine refers, namely, that on "The Changes of Temperature produced by the Rarefaction and Condensation of Air," but were published previously, and had not been seen by Rankine.

From this mechanical hypothesis of heat and the constitution of matter, Rankine, by his remarkable innate mathematical facility, developes as general equations the relations between pressure, volume, temperature and heat, of the matter so constituted. Using the "Experimental Results of Regnault," published in 1847 ; the experiments of De la Roche and Bérard, as to specific heat, together with the velocity of sound, and Laplace's extension of Newton's theory, as data from which to determine the values of his symbols, Rankine at once applies the theory to determine relations hitherto not experimentally determined, or not determined with sufficient accuracy, for air and steam, and this with great success.

These equations of Rankine, if they truly express the thermodynamical condition which holds in any particular mechanical arrangement, must implicitly conform to the *general laws*, which hold for all arrangements of matter ; but they obscure rather than indicate the generality or importance of such laws. Based on a definite hypothetical mechanical constitution of matter, they implicitly conformed to the law of the conversion of that form of energy named heat to that called mechanical energy, but they did not explicitly call attention to the general importance of this law, or to the importance of the mechanical equivalent of heat. In the same way Rankine's equations conformed to Carnot's law, but in no way indicated its generality or importance. Thus it was that Rankine, having failed to recognize the general truth of Carnot's law, did not know that it was to be obtained from his equations until he looked for it after its importance had become known to him.

To the importance of Joule's mechanical equivalent of heat, Rankine's attention had already been called by Joule's papers. This Rankine acknowledges in his first paper and, at the same time, declares his conviction of the unexceptional character of the evidence adduced by Joule of the convertibility of heat and mechanical effect; but, at the same time, declares himself dissatisfied with the value of the equivalent as determined by Joule, on account of the smallness of the differences of temperature measured in Joule's experiments, and the numerous possible sources of error to which such experiments are necessarily exposed, which sources he points out in a somewhat patronizing manner.

Rankine then proceeds :—

" The best means of determining the mechanical equivalent of heat are furnished by those experiments in which no machinery is employed. Of this kind are experiments on the velocity of sound in air and other gases, which, according to the received and well-known theory of Laplace, is accelerated by the heat developed by the compression of the medium."

There is a speciousness in this argument arising from the fact that these experiments on sound only become available for the purpose when the true value of at least one of the specific heats of air is known, and that the determination of this involves " machinery," and is a matter of extreme difficulty. Rankine, in the natural anxiety to render his theory, which he had considered in 1842, independent of Joule's subsequent work, for the moment overlooked this speciousness, and in his first paper deduces, from the data already mentioned, the mechanical equivalent of heat to be 695·6, whereupon he remarks, " I have already pointed out the causes which tend to make the apparent value of the mechanical equivalent of heat, in Mr. Joule's experiments, greater than the true value. The difference between the result I have just stated and those at which he arrived do not seem greater than those causes are capable of producing when combined with the uncertainty of experiments, like those of Mr. Joule, on extremely small variations of temperature."

That these criticisms of Rankine's should, when he first saw them, have been painful to Joule, there is no wonder; at the same time he must (knowing as he did the insufficiency of Rankine's data; having himself previously followed the same determination by Laplace's theory), have anticipated the result. This followed the same year. Rankine's paper was read, February, 1850; on December 2nd, another paper of Rankine's was read, in which he explains that since his reflections on the accuracy of Joule's experiment, he has seen the detailed account of Mr. Joule's last experiments in the *Philosophical Transactions* for 1850, which, he continues, "have convinced me that the uncertainty arising from the smallness of the elevations of temperature is removed . . . . . and that the necessary conclusion is, that the dynamical value assigned by Mr. Joule to the specific heat of liquid water, viz., 772 feet per degree of Fahrenheit, does not err by more than two or at most three feet, and that, therefore, the discrepancy originates chiefly in the experiments of De la Roche and Bérard."

"I therefore take the earliest opportunity of correcting such of my calculations as require it so as to correspond with Mr. Joule's equivalent."

There is no doubt that whatever was lost in the generality of Rankine's mechanical theory of heat from its hypothetical foundation, that the definition and suggestiveness arising from this foundation was of the greatest advantage in its own mathematical development as well as in its application; and that this development and application were of the greatest use in directing the development of the Theory of Thermo-dynamics founded on the perfectly general laws of Joule and Carnot, which followed immediately after. But Rankine was not alone in affording such assistance. In 1848, Joule had, from the development of his own theory and its application to bring Carnot's principle into accord with his experiments on air, been led to the discovery of the true form of the relation discovered but not defined by Carnot, between the proportion of heat converted into work and the initial and final temperatures at which the heat existed. All Carnot had been able to show was that the work done was dependent on these temperatures only, in a form that was expressible by $C(T_1 - T_2)$ where C might depend in any unknown way on the temperatures. In a letter, dated December 9th, 1848, to Sir William Thomson, Joule suggested that the value of C (Carnot's function) was the quotient of the heat received at the temperature T_1 divided by the temperature at which the heat was received measured from the absolute zero as discovered by Joule.

In February, 1850, there was read to the Berlin Academy of Sciences, a paper on "The Mechanical Action of Heat," by Rudolf Julius E. Clausius. This paper, read almost simultaneously with Rankine's, contained almost an identical theory, and its almost identical application. Clausius' theory had a somewhat different foundation however. It was primarily founded on Joule's law, and a general hypothesis as to the dynamical constitution of gas, from which Clausius, like Joule before him, obtained the absolute zero of temperature, and the same definite form of Carnot's function. Clausius then, for the first time, so expressed Carnot's theory as to render its foundation consistent with Joule's law. This partly hypothetical, partly general, foundation led Clausius, in the mathematical development of his mechanical theory of heat and its application almost, though not quite, as far as Rankine had gone.

These theories, coming as it were as "bolts from the blue," caught Sir William Thomson still endeavouring to evolve the perfectly general foundation of thermo-dynamics, from Joule's discovery and Carnot's theorem. Rankine's was the first, as after being read February 4th, 1850, Rankine's paper was referred by the Royal Society of Edinburgh to Thomson to report upon. This Sir William Thomson did, suggesting certain improvements which the author acknowledges and adopts, and for which he expresses his thanks. The hypothetical foundation of Rankine's theory (though exciting Sir William Thomson's admiration no less than the mathematical development erected on this foundation, and the novelty and importance of some of the results, indicated in Rankine's initial application of it) afforded Sir William no direct assistance in finding the general foundation for which he was seeking; though it could not but have confirmed him in his view as to the possibility of such a foundation. Clausius' paper, published in Pogendorff's *Annalen* for March and April, 1850, reached Sir William Thomson early in 1851, just after he had succeeded in proving Carnot's theorem on a foundation consistent with Joule's discovery; having previously determined the general equations of the mathematical theory of heat founded on the two principles—Joule's and Carnot's—and applied the equation to Regnault's observations. Sir William Thomson

K

was thus forestalled by Rankine in the results to be obtained by the development of the mathematical theory of heat, and by Clausius in showing that Carnot's theorem was reconcilable with Joule's discovery. Owing, however, to the hypotheses involved in both Rankine's and Clausius' theories, it was left to Sir William Thomson to clear these hypotheses away, and show that Joule's discovery of the equivalence and convertibility of heat and "work," together with Carnot's theorem (re-founded on the axiom first discovered by Clausius) constituted, alone, a sufficient foundation for a complete theory of thermodynamics, explaining all the known physical phenomena and their quantitative relations. This Sir William Thomson immediately did. In March, 1851, a paper "On the Dynamical Theory of Heat, with Numerical Results from Mr. Joule's Equivalent of a Thermal Unit, and M. Regnault's Observations on Steam," by Sir William Thomson, was read to the Royal Society of Edinburgh. In this paper, which, in virtue of the absolute generality of his foundation, and the pre-eminence of his powers in mathematical analysis and sympathetic expression, easily placed the author in front of his two contemporaries as an exponent of Thermodynamics, Sir William Thomson gives all credit to Rankine and Clausius, and expresses himself with enthusiasm as to their work; while the paper is one continuous eulogy of the work of Joule, not only in discovering the mechanical equivalent but throughout its entire range. He makes reference to and gives quotations from all Joule's papers, from the first presented to the Royal Society in 1840, "On the Production of Heat by Voltaic Electricity," continually expressing his admiration for the conclusiveness of Joule's proofs and the philosophical acumen of his deductions, no less than for the beauty of his methods and the unequalled accuracy of his experiments.

Such a recognition, coming after a careful and critical consideration over a period of four years from such a source, was well worth Joule's patience in waiting for it. It not only assured him the full apprehension and appreciation of his work, and what was yet more dear to him, the truth of his philosophical deductions, but sounded his trumpet with a blast which, while delightfully pleasant at the time, was yet strong enough to awaken the scientific sleepers all over the world, and go on echoing into succeeding generations.

In the meantime, Joule had been showing the importance which he attached to the mechanical equivalent of heat by a further investigation, into which he had thrown his whole power; and, aided by his experience and the perfection of his appliances, had made the final determination.

The paper containing this was communicated to the Royal Society by Faraday, in June, 1849, before the publication of Rankine's paper. It was accepted and published in the *Philosophical Transactions*, 1850. The research contains only one new departure from those previously published, which was the determination of the equivalent from experiments on the friction of cast-iron. But, in that the experiments on fluid friction with water and with mercury, recounted in this paper, greatly exceeded in number those previously given, being 41 with water, 52 with mercury, and that they were made with more perfect apparatus, the agreement between the results so obtained was closer; besides being confirmed by the experiments with cast-iron. These led Joule to the final conclusions;—He says: " I consider that 772·692, the equivalent derived from the friction of water, is the most correct, both on account of the number of experiments tried and the great capacity of the apparatus for heat, and since, even in the friction of fluids, it is impossible to avoid vibration, and the production of slight sound, it is probable that the above number is slightly in excess. I will, therefore, conclude by considering it as demonstrated by the experiments contained in this paper,—

1st. That the quantity of heat produced by bodies, whether solid or liquid, is always proportional to the quantity of force expended.

2nd. That the quantity of heat capable of increasing the temperature of 1lb. of water (between 80° and 60°) by 1° Fahrenheit, requires for its evolution the expenditure of a mechanical force represented by the fall of 772lb. through one foot."

This was Joule's final result, and is still accepted, as expressing the true value within the limits of accuracy of any known means of determination; while it now enters into almost all physical calculations as well as those which guide the practical conversion of heat into "work."

The paper in which this result is given contains a very interesting preamble, in which Joule gives a full sketch of the mechanical theory of heat, commencing with Count Rumford.

In dealing with Rumford, he does what Rumford had not done, and what had not been done in the meantime. By applying to Rumford's definite statement of the heat generated in 2h. 30m. (26·58lbs. of water raised 180° Fahrenheit) by the "work" that could have been done by one horse, the recognised measure of 1 horse power, 33,000lbs. raised one foot per minute, he shows that the experiments of Rumford gave an approximate equivalent of 1034 foot lbs. per 1° of heat. He then describes Sir Humphrey Davy's experiments of rubbing two pieces of ice together until they were melted. He also mentions Dulong's statement that equal "volumes of the same elastic fluids, taken at the same temperature and under the same pressure, being compressed or dilated suddenly, disengage or absorb the same *absolute quantity of heat*." He then mentions the works of Faraday, Grove, and Mayer, as preceding his own early work in advancing the idea that the so-called imponderable bodies are merely the exponents of different forms of force. He mentions the experiments of H. Seguin on steam, as affording an approximate determination of the mechanical equivalent, and finally says:—

"The first mention, so far as I am aware, of experiments in which the evolution of heat from fluid friction is asserted, was in 1842, by M. Mayer, who states that he has raised the temperature of water from 12° C. to 13° C., by agitating it, without, however, indicating the quantity of force employed, or the precautions taken, to secure the correct result."

In this historic sketch, the measure of credit Joule gives to the importance of the discoveries of his predecessors is in no sense minimised or given grudgingly, but, on the other hand, is ample (containing references to their work), and only to be justified after viewing their results by the better light of his own. He clearly does not view them as competitors for fame, but as fellow-labourers in the task of convincing the scientific world of its error, who afford powerful advocacy to views he himself is advocating.

It must, also, be doubted whether, but for this sketch, the work of his immediate predecessors, Seguin and Mayer, would have been recognised ; how Joule discovered it is not clear ; it can only have become known to him about the time he was writing. Neither is it likely that a recollection of the views and experiments of Rumford and Davy, which was quite lost sight of at the time, would have immediately revived had it not been for Joule's sketch.

It was, no doubt, this sketch that furnished the means by which an attempt was made to diminish the growing importance that was being attached to Joule's work in its effect on the revolution in philosophic thought, by setting up the work of Mayer, in particular, as diminishing the lustre of that of Joule. Although, in the controversy which this attempt succeeded in raising, in their warmth, some of Joule's supporters somewhat minimised the credit due to Mayer, Joule himself took no part in this, but always insisted on giving Mayer and others full credit for what they had done, declaring himself more than satisfied with the share which he received.

CHAPTER IX.

MIDDLE LIFE. — *Summer of his Life.—Acton Square.— Welcomes Mathematical Assistance.—Comparative Rest.— Amalgams.—Air Engine.—Joint Research with Thomson. — Vice - President of the Society. — Visit of Sir William Thomson. — Birth of his Daughter. — Royal Medal. — Death of his Wife. — Return to Oakfield.— Honours.—Electrical Welding.—Joint Research Resumed. —Memoir of Sturgeon.—Thermo-dynamical Properties of Solids.—Council of Royal Society.—Railway Accident.— Work at Oakfield.--Thorncliffe.--His Experiments Stopped. —President of the Society.—Honours.— Visits.—Small Scale Researches.—Determination of Mechanical Equivalent of Heat from the Thermal Effects of Electricity.— Propagation of Joule's Views.—Hirn's Verifications.*

The end of 1849 finds Joule, then 31 years of age, already commencing the short summer of his life. His regular attendance at the meetings of the British Association since 1842, and the office of secretary to the Chemical Section, to which he was elected at the meeting in Cambridge in 1845, have secured him a large acquaintance amongst the leaders of science who attended these meetings. He has also made the acquaintance of Professor Bunsen, who called on him at Oakfield with Sir Lyon Playfair in 1845, and he has paid several visits to London to see Sir Lyon Playfair, and Faraday, or to stay with his former tutor, Mr. Tappenden. He is Corresponding Member of the Royal Academy of Sciences, Turin, to which he was elected in 1848. Since April 9, 1849, he has been established in No. 1, Acton Square, Salford, where he resides with his wife (and where, on May the 18th, 1850, his son Benjamin Arthur was born). He is to be elected Fellow of the Royal Society in June, 1850,—Sir William Thomson receiving the like honour in 1851, and Rankine in 1852. As Secretary of the Society since 1846 he has already an acknowledged and important position as a citizen of Manchester, which insures him contact with all those who take interest in literature and science and friendly intercourse with the eminent men on the Council.

By his final determination of the mechanical equivalent of heat, in 1849, he finds himself released from the constraint in which his own instinctive perseverance has held him, causing him to pursue to the end the clue which arose from his boyish attempt, in 1838, to improve Sturgeon's electromagnetic engine. He has the assurance of his own philosophical insight that his discoveries reveal fundamental truths of the greatest importance, not only in philosophy, but to the future material well-being of mankind. And he finds that the clouds which have so far enveloped his work are lifting, and that the sunshine of the highest scientific as well as friendly sympathy and recognition has already begun to break in upon him. Just as his solitary task of exploring and paving the way, which by its exigences in engrossing his attention has prevented him from feeling, though not from perceiving, how solitary he has been, is completed, and his mind is free, he finds friends already using the main road, acknowledging its importance and constructing branches.

Had the clouds not lifted—had Joule still found his work unrecognised—he might have felt himself, however ill-qualified for the task, compelled to continue, the development he had already commenced of the several mathematical sciences which directly followed his discovery of the definite relations between the physical and mechanical sources of heat. But the assurance that such mathematical powers as were evinced by Sir William Thomson, Rankine, and Clausius, were inevitably directed to the subject, besides the fact that the work itself was already in great part accomplished by the two latter, before Joule had become aware of their assistance, not only relieved him from all necessity to continue it, but in a great measure took it out of his hands, and carried it into a region where the language was such that Joule, with his scant mathematical education, was ill-qualified to follow, much less to lead the advance.

That Joule resented this does not appear in the slightest, rather that he welcomed it with enthusiasm. He evinced an admiration which almost amounted to reverence for the mathematical powers displayed, himself looking on until demand was made for further experimental work on incidental points, when he again lent his unequalled powers to the mathematicians.

For the next two years, while Thomson, Rankine, and Clausius were performing mathematical feats, which not only secured them the front place amongst the mathematicians of their time, but raised the Universities of Glasgow and Zurich above all others as seats of the most advanced mathematical and physical philosophy, Joule had little to do but enjoy the rest he had earned, and the gratification which he must have felt in the chivalrous and enthusiastic honour which these men paid to his work.

The sensations consequent on having discovered and accurately determined a physical constant of fundamental importance have been experienced by few, and Joule was probably the first who, in addition to these sensations, experienced those which must have arisen in finding that his contemporaries who were carrying on his work (Clausius and Rankine) eagerly agreed in the suggestion of Sir William Thomson to use J as the mathematical symbol to signify the Mechanical Equivalent of Heat, and so to associate for all time Joule's name with his discovery.

Although Joule seems to have done little more than amuse himself during 1850 and 1851, he was not idle. Having reached the terminus of the main line of his inquiry, and finding the immediate branches occupied by those better qualified than himself to prosecute them, he appears to have taken up, for one thing, a new line of molecular study requiring experimental investigation, that of "Amalgams," on which he made researches at his house, Acton Square, in 1850, publishing the results partly at the British Association in that year, and subsequently in 1862 in Memoirs of this Society. At the same time, however, he was continuing in his laboratory at Oakfield his investigations on Electro-Magnets, which he published in two papers in the *Philosophical Magazine*, 1852.

At the same time Joule, having proved the impossibility of exceeding the economy of the steam engine by means of electro-magnetic engines, showed that he retained his early ambition to accomplish a useful practical application of his researches, by undertaking the invention of a hot air engine which should work under conditions allowing of greater ranges of temperature than are obtainable in the steam engine, and so, according to the new development of

Thermo-dynamics, obtain greater economy. A description of this invention, with a discussion of its possible economic advantages as shown by the new theory, was read before the Royal Society, June, 1851, and published in the *Phil. Trans.*, 1851.

Joule did not construct an engine or undertake any experimental investigation connected with it; but merely described in detail his engine, and gave the mathematical theory of its action, in which he acknowledges the assistance of Sir William Thomson. Under the name of "Joule's Air Engine," the theory of this engine is included in almost all text books on Thermo-dynamics. Joule's air engines have been tried over again, and although, like all hot air engines, have (from incidental causes not to be realized without experience) failed to rival the steam engine in general economy or usefulness, a number of Joule's engines have been usefully employed for special purposes. The paper is, however, chiefly interesting as showing that in the early stages of the mathematical development of the theory of Thermo-dynamics, Joule followed until he perfectly realized the general significance of the theory as depending on Carnot's theory as well as his own. Beyond this Joule does not appear to have gone.

From his attempts to start incidental lines of experimental research, Joule was diverted in 1852 by the requirements that became apparent for further experiments to determine the data to complete the application of Thermo-dynamics to gases, and to verify the conclusions arrived at by those who were developing the results of this theory. This led him to undertake a joint investigation with Sir William Thomson, in 1852, which was continued, with interruptions, till 1859.

To the reprint in 1885 of the papers resulting from this investigation, Joule prefixes the note containing the reference to his meeting with Sir William Thomson, at Oxford, already quoted, Chapter VIII.; this note continues :—

"My work with Thomson was chiefly experimental, performed in Manchester and the neighbourhood. We pursued the discussion of the thermal effects of fluids in motion until the experiments were interrupted by the action of the owners of the adjacent property, who, on the strength of an obsolete clause in the deed of conveyance, threatened legal proceedings, the cost of which I did not feel disposed to incur."

This research was directed, in the first place, to verify a conclusion resulting from the application of Thermo-dynamics to Regnault's results.

The researches of Regnault, published in 1847, while they confirmed Boyle's law, that the density of air at constant temperature increases in the direct ratio of the pressure, to a much greater degree of accuracy than was previously obtained, showed, nevertheless, that the density increased slightly faster than the pressure, and that this increase was still more marked in the case of carbonic acid.

According to Thermo-dynamics, this discrepancy suggested that there would be a similarly small cooling effect when gas is forced through a small aperture and allowed to come to rest. Joule had shown that this cooling was insensibly small in his experiments on air; but this was quite consistent with the very small deviation from Boyles' law discovered by Regnault, so that to find the cooling effect it was necessary to make experiments more directly adapted to the elucidation of this point. A method suggested by Thomson, that of forcing air along a pipe with a very narrow neck in it, and measuring the temperature on each side of the neck, was adopted, and after two years' work, gave a cooling effect perfectly consistent with the deviation from Boyle's law for air, hydrogen, and carbonic acid.

The experimental research was very severe, and involved much larger apparatus than anything Joule had previously used, the final results being obtained in the Salford Brewery, only by the aid of a steam engine developing three horse-power, and even then the greatest differences of temperature with air were very small.

As both the research and the method employed were suggested by Sir William Thomson, it is, here, chiefly interesting as showing how Joule was occupied during the years 1852 and 1853, and as affording an instance of the unique powers he displayed in thermal mensuration.

In June, 1850, Joule visited Mr. Tappenden in London; and was admitted a Fellow of the Royal Society by the Earl of Rosse, being introduced by Dr. Forbes. In July, with his wife, baby, nurse, and sister, he went to Edinburgh to the British Association.

Joule held the office of Secretary in this Society from 1846 to 1850. His colleague during the last three years was Mr. E. W. Binney, with whom the friendship thus begun became very intimate, and lasted to Mr. Binney's death in 1881. In 1850 Joule vacated the office of Secretary on being elected one of the four Vice-Presidents, Sir William Fairbairn being another Vice-President, and Eaton Hodgkinson President. The interest Joule took in the work of the Society during these years was a subject on which Mr. Binney often remarked in later years.

On the 14th October, 1851, he was in London taking part in the closing ceremony at the Exhibition, and on the 17th December, paid his first visit to Sir William Thomson, returning on the 24th. In May, 1852, Sir William Thomson was visiting Joule at Acton Square; and on the 17th inst, with Joule and his brother Benjamin, went by rail to Huddersfield, and thence to Holmfirth, to see the devastation caused by the bursting of the reservoir. Of this excursion Mr. J. St. B. Joule writes, "Although a great amount of straightening up had been accomplished, we saw enough during our walk to the Bilberry Reservoir to form a good idea of the immense destruction the bursting had caused. Steam engines demolished, factories cut in two—the half gone—and, having carefully inspected the part of the embankment which appeared to have first given way, we (all three being good walkers) resolved to strike over the hills, first taking the Hocking Holmes Moor, until we could reach the M. S. & L. Railway for Manchester, and after a long walk we came down near the line some three miles below Woodhead."

On June 13, 1852, Joule's daughter, Alice Amelia, was born; she was christened July 7th, Sir William Thomson being godfather.

On November 29th, Joule attended the Anniversary Meeting of the Royal Society, to receive the Royal Medal which was awarded to him; Huxley receiving one at the same time, and Darwin another the next year.

On June 8th, 1854, Joule's second son was born, but only lived till the 28th.

In the death of his wife, which occurred September 6th, 1854, leaving him with one son and one daughter, Joule suffered a loss, which not only disturbed his immediate career, but which, by confirming his naturally retiring disposition, deterred him from taking the prominent place in the republic of science, which the recognition of his work was now beginning to thrust upon him.

Holding no office or post which necessitated his exerting himself, he had only his experimental work to divert his attention from his loss, and this he seems to have been unable to pursue for more than a year. He gave up his house in Acton Square, and returned with his children to his father's house, Oakfield, on October 19th, to reside. At his father's house he had the society of his brothers. Writing to the author, Mr. B. St. J. B. Joule says: "The death of his wife affected my brother very much, but when he had left the house and rejoined us at Oakfield, his previous condition was pretty well regained." Greatly as this complete retirement of Joule was to be regretted, it could not prevent the rapid growth of his fame.

In 1855 Joule's discoveries were already before the elect of science. Rankine, Thomson, and Clausius had virtually completed their development of the mathematical theory of Thermo-dynamics and its application to the principal problems in physics and heat engines, and their papers, together with Joule's, in the transactions of the several societies, were published all over the world. Rankine's article "Heat, the Mechanical Action of," in Nichols' *Encyclopædia*, the first formal treatise on Thermo-dynamics, was already written. Clausius and Boltzman had commenced the development of the modern dynamical theory of gases from Joule's extension of Herapath's theory.

Honours are now beginning to fall thick on Joule, who was elected Hon. Member of the Cambridge Philosophical

Society, 1855; LL.D. of Trinity College, Dublin, 1857, in which year he was also elected on the Council of the Royal Society; Hon. Member of the Inst. of Engineers and Ship Builders of Scotland, 1859; and of the Philosophical Societies of Glasgow and Basle, 1860; also D.C.L. of Oxford in the same year.

In 1855 Joule also invented the now well-known process of welding metals by means of the electric current, and succeeded in proving its practicability early in 1856, the first experiment being made "in the Laboratory of Professor Thomson, in Glasgow." "More recently," he says, in the paper read to this Society, March 4th, 1856, "I have made several experiments in which the wires were placed in glass tubes, surrounded with charcoal, &c. With a battery of six Daniell's cells I have then succeeded in fusing several steel wires into one, uniting steel with brass, platinum with iron, &c. I have no doubt the process would advantageously supersede that of soldering, &c." He then proceeds to determine the consumption of zinc in a Daniell's battery required for the process, and finds it to be as low as three-quarters of a pound of zinc to fuse one pound of iron.

In the summer of 1855, at the kindly instance of Sir William Thomson, Joule resumed his experimental work on electro-magnets, with the immediate object of ascertaining whether the results obtained by Jacobi and Lenz, as to the relations between the magnetism developed, the dimensions of the iron, the coil of wire, and the current, were not, with the apparently discordant results which Joule had previously found, particular cases of the general law which had been suggested by Sir William Thomson.

In 1856 they again resumed their experiments on the thermal effects of fluids in motion—this time directed to determine the temperature of a body moving through the air, and so to prove the assumption involved in Joule's explanation of shooting stars, viz., that a body moving through the air would be raised to a temperature greater than that of the air through which it is moving. Although this assumption is in strict accordance with thermo-dynamics, and as applied to such velocities as those of meteorites, may pass without appearing contrary to experience; yet to say that a body placed in a high wind, or whirled through still air, will have its temperature raised above that of the air, does seem paradoxical and against ordinary experience. The research, however, which was continued till the end of 1859, completely confirmed the assumption, a thermometer moving at 100 miles an hour being raised some 2° Fah., and the temperature increasing with the square of the velocity; so that this rise was quite sufficient to account for the meteoric effects of bodies moving at 19 miles a second.

These investigations appear to be the only experimental work undertaken by Joule between 1854 and 1857; but, in 1856, he wrote the account of Sturgeon, published in the Society's *Memoirs*, Vol. XIV., second series. In this "Memoir" Joule reproduces the vindication which he had previously written of Sturgeon's claim to be the discoverer of the *soft iron electro-magnet* and *electro-magnetic engine*, and the "*commutator*."

In 1857, Joule had commenced his very important researches "On the Thermal effects of stretching and compressing Solids," "On compressing Liquids," and "On some Thermo-dynamic Properties of Solids."

These researches, though conducted by Joule, appear to have been largely suggested by Sir William Thomson's theoretical anticipation of results not previously observed to exist, or only incidentally observed; as in the case of india-rubber, which, when placed between the lips and stretched, produces a sensation of warmth, a phenomenon commented on by Mr. Gough in the Society's *Memoirs*, Vol. I., second series.

Sir William Thomson had deduced from the general theory of thermo-dynamics, that when an elastic body has its form changed by definite forces the heat which becomes latent is the product of the absolute temperature multiplied by the forces, multiplied again by the change of dimensions in the direction in which the force acts, caused by raising the temperature of the body one degree, whence, knowing the specific heats, weight, and coefficient of expansion, the alteration of temperature consequent on subjecting a solid or liquid to stress could be quantitatively found.

Joule verified the truth of this theory in the cases of iron, hard steel, cast iron, copper, lead, gutta-percha, vulcanized india-rubber, various kinds of wood, and water.

At the conclusion of his paper "On some Thermo-dynamic properties of solids," *Phil. Trans*, 1859, Joule, referring to his last determination, in which he had measured the heating and cooling effect in a spiral spring to be $0°·00306$ Fah., while the theoretical effects to be expected were $0°·00406$, remarks :—

"Thus even in the above delicate case is the formula of Professor Thomson, completely verified. The mathematical investigation of the thermo-elastic qualities of metals has enabled my illustrious friend to predict with certainty a whole class of highly interesting phenomena. To him specially do we owe the important advance which has recently made a new era in the history of science, when the famous philosophical system of Bacon will be to a certain extent superseded, and when instead of arriving at a discovery by induction from experiment, we shall obtain our largest accession of new facts by reasoning deductively from fundamental principles."

Joule was in London in 1855, and again early in 1856. On May 25, 1857, he had another visit from Sir William; and on the 30th, Joule and his brother went with Sir William and Mr. Crum to Liverpool to inspect the Atlantic cable.

In November, 1857, Joule was elected on the Council of the Royal Society, which caused him to make frequent visits to London.

On February 3, 1858, he was invited by Scott Russell to inspect Brunel's plan for launching the Great Eastern, when he had lunch with the Duke of Argyll and his little son, the Marquis of Lorne; and on his return wrote a letter in the *Courier* (January 22) on "A Stranded Leviathan."

He was at Glasgow, February 27, and at Peterborough, March 15, observing the annular eclipse of the sun. On May 6, he was again in London, returning on the 10th, when he encountered an incident, of which his brother gives the following account:—

"10th May, 1858. I went to London Road Station to meet James. I had long to wait. He left London by the London and North-Western train at 9 a.m. Near Nuneaton he noticed a rapid diminution of speed, which was quickly followed by a crash, and his carriage went over on one side. He scrambled out and found that the carriage, which on starting was fourth from the engine, was now the first. The accident was caused by a cow which had strayed on the line. The engine which remained on the line was covered by the remains of the cow. It was supposed that if the full speed had been maintained the engine would have crushed the obstacle, a large bone, which had arrested its progress. James was astonished when he saw the engine men eating their dinner with as much *sang-froid* as though nothing had happened, while the passengers were in a state of the utmost terror. An officer in charge of soldiers was so much excited, that he was brandishing his sword in an adjacent field. The carriage which James had occupied was tumbled down the embankment to get it out of the way. Three persons were killed. James was studying a mathematical book at the time, and when we arrived at home we found the pages which he had open were covered with pulverized glass."

This accident made Joule very nervous in travelling by railway for a long time, and in consequence he declined to be again nominated as a member of the Council of the Royal Society. He was in the Isle of Man in the summer, and at the British Association in Leeds, in September. On December 30th, Joule's father died, having been unwell for some years.

At this time Joule was hard at work in his laboratory at Oakfield. The Brewery in Salford, where he, with Sir William Thomson, had made the experiments on the thermal effects of fluid in motion in 1853, being sold in 1854, he had his steam engine and appliances removed to Oakfield, and what could not be placed in the laboratory, he worked with in the open. Of Joule's work, Mr. B. St. J. B. Joule writes: "My brother was at Whalley Range (Oakfield) very busy with his experiments, many of which were decidedly dangerous owing to the pressure he made use of. During this period, for some months he could not find time to take his meals properly—just ran in and out again. The experiments were so delicate that many were carried out in the night, because a cab or cart passing along the road disturbed them, though the laboratory was at the back of the stables."

"The conduct of Thomas Woods, M.D., in 1860, gave him much trouble; and he was very angry with him for suggesting that he had made alteration in or addition to the essay he sent to Paris, when it was published in England, but the deposition of the translator, Mr. L. A. J. Mordacque, to the effect that the published paper was precisely word for word the same as that he had translated, settled that matter. Dr. Woods concluded the controversy in truly Irish fashion, by saying Dr. Joule had made so many discoveries that he thought he might have allowed him two or three."

In 1860 and 1861 Joule was much engaged experimenting on the condensation of steam, and published his well-known results on the surface condensation of steam in the *Philosophical Transactions* for 1861.

In this year his experimental work was subject to the stoppage already referred to. This happened as follows:—

After the sale of the Brewery in 1854, and the removal of his steam engine to Oakfield, his father, having been unwell for some years, died in 1858, leaving his house to be sold.

In consequence of this, Joule, in 1861, having purchased "Thorncliffe," Old Trafford, removed to that house, with his apparatus, including the steam engine. It was then Joule's work was stopped in a way which caused him great and lasting disappointment. His brother writes: "What really affected him was the refusal to be allowed the use of his one-horse power steam engine, which would have been allowed but for the perverse opposition of . . . . who resided at Old Trafford, and, in spite of a memorial from the other residents, insisted upon an item in the deed which prohibited any steam engine being used. My brother was anticipating a series of important experiments in conjunction with W. Thomson, for which a grant had been obtained. However, he made some experiments, *e.g.*, his sensitive barometer, using the coach-house for the purpose. He took the first opportunity of selling his house."

In 1861 what had been the general wish of this Society for some years, was gratified by the election of Joule to the office of President, in succession to Sir William Fairbairn, who vacated the chair, after having occupied it five years, in consequence of the adoption in the meantime of a rule that the president should retire after two years of office. Sir William Fairbairn, Angus Smith, and E. W. Binney were elected Vice-Presidents at the same time, and Drs. Schunck and Roscoe (Sir Henry) were Secretaries; and after this time Joule was elected President on every possible occasion as long as his health would allow him to accept the office.

After his removal to Thorncliffe, and the stoppage of his experimental work, Joule seems to have spent much more time away from home. Thus, in May, 1861, he took his family and sister to the Isle of Man, where his younger brother was living. In September he spent a fortnight with his elder brother in the North-west of Ireland, and in October was in London for a week. In 1862 he made three visits to the Isle of Man, besides spending a month in Wales and some time in London and Cambridge. In 1863 he was frequently at Douglas, staying with his younger brother whose health caused much anxiety, and whose funeral Joule attended, March 3rd, 1864. Joule and his family spent July at Mr. Tuppenden's, Eatton Court, Turnham Green. In January, 1865, he spent a week at Greenock, during which he delivered an address on the occasion of the opening of the Watt monument. In the summer of 1866 he and his family were at Portrush, and in the autumn of 1867 at Port Erin.

In 1868, Joule, having sold his house in Old Trafford, removed, 16th October, to No. 5, Cliff Point, Lower Broughton, Manchester.

In 1861, Joule was a Vice-President of the British Association, in 1864, Member of the Philosophical Society of Halle, and in 1867, Fellow of the Royal Society of Edinburgh.

In 1864, Joule's portrait, by G. Patten, from which the engraving accompanying this memoir has been taken, was presented to the Society by E. W. Binney and other members.

Although in 1861 Joule was prevented continuing the experiments on steam, or requiring steam, he still managed to do a great deal of experimental work, mostly the improvement of thermometers, barometers, instruments for measuring magnetic declination, and horizontal magnetic intensity, and tangent galvanometers. On these subjects, and on observations of certain natural phenomena, he communicates twenty papers to this Society between 1859 and 1868.

In 1867 he makes experiments for the "Determination of the Mechanical Equivalent of Heat from the Thermal Effects of the Electric Currents," for the Committee on Standards of the British Association. What Joule undertook to determine was, not the Mechanical Equivalent of Heat, but, the heat produced by a definite current in a definite time when traversing a conductor of definite resistance. This was comparatively a simple matter, and he was in no way responsible for the mechanical value assigned to the electric action as measured by the British Association unit of resistance; the value of which unit was determined by others. Joule accomplished his part with his usual skill and accuracy, devising an absolute current meter for the purpose. Then, dividing the mechanical value of the electric action (obtained by the B.A. unit) by the heat, he obtained 782 lbs., which, although about one per cent. larger than the value obtained with water, was a splendid verification of the electric measurements. Nevertheless the discrepancy, until it was explained, cast a possibility of doubt on the 772 lbs.

After the mathematical development of the general theory of thermo-dynamics had been completed by Thomson, Rankine, and Clausius, in 1855, the interest in the subject was not allowed to lapse, but was rather enhanced by the less general direction which was given to it, in the development of the dynamical theories of gases by Clausius, and of electricity by Thomson and Helmholtz, whose work was so ably taken up by C. Maxwell and P. G. Tait and Boltzman, and secured the attention of all the leading physicists of the time.

Rankine's *Manual* on "The Steam Engine," published in 1860, and already in 1868 at the end of the third edition, had brought the practical importance of the new theory before the engineering profession, pointing out clearly and definitely the economic possibilities, as well as the advances in mechanical construction necessary for this realisation.

At the same time Hirn had lent his powerful aid to propagate Joule's views and the new theory on the Continent. In 1865, Hirn repeated Joule's experiments for the mechanical equivalent of heat by the friction of water; introducing a great simplification in the measurement of the work expended. This was accomplished by Hirn's recognition of the principle that the moment of resistance which the water opposes to turning of the paddle must be equal to the moment of resistance which the containing vessel opposes to the water, so that instead of measuring the force spent on the moving paddle it was only necessary to measure the force to keep the can from turning. Hirn also succeeded in verifying, for the first time, Joule's views expressed in 1847 as to the conversion of heat into work in the steam engine, as shown by the difference between the heat received from the boiler and discharged by the condenser, thus instituting that system of steam engine trials which now govern all practice. At this time, too, the works of John Tyndall and of Balfour Stewart had gone far to render Joule's views popular and introduce them into the curricula of school education.

CHAPTER X.

LATER LIFE.—*Joule in 1869.—Estimation in the Society—Sociability—Copley Medal.—Institute of France.—Numerous Communications.—Display of Character.—Alleged Effect of Frost.— Performance of Electro-Magnetic Engines.—President of British Association.—First Failure of Health.—Verification of Final Determination of Mechanical Equivalent of Heat.—Change of Residence ; 12, Wardle Road, Sale.—Royal Pension.—Honours.—Collected Scientific Papers by the Physical Society of London.—Declines to be President of British Association, 1887.—Failing Health.—Death.—Memorial Statue in Manchester.—Memorial Tablet in Westminster Abbey.—International Memorial.—Portraits of Joule.*

It was at a meeting of the Society in 1869 that the author first saw Dr. Joule, who was then in the chair. Although having of necessity become imbued with the transcendent importance of Joule's work in physical Philosophy, and the appreciation in which this was held, and regarding him much in the same light as we regard Galileo or Newton, as a being of another order, the impression on first sight contained no suggestion of disappointment. That Joule, who was then 51 years of age, was rather under the medium height; that he was somewhat stout and rounded in figure ; that his dress, though neat, was commonplace in the extreme, and that his attitude and movements were possessed of no natural grace, while his manner was somewhat nervous, and he possessed no great facility of speech, altogether conveyed an impression of the simplicity and utter absence of all affectation which had characterised his life; while his fine head and the reflective intelligence of his grave face accorded with the possession and long exercise of the highest philosophical powers. Another thing, too, calculated to impress a new member, was the obvious respect, amounting to veneration, that was plainly evinced by all the members present at the meeting, as was also the kindly and encouraging remarks which Joule, as president, made opportunity to address to the new member, either during the meeting or after it was over.

Such were the first impressions, which were only strengthened by further and closer intercourse, extending over seventeen years. It soon became evident that it was not merely veneration arising from the fame of Joule that inspired the members of the Society, but that it was an attachment arising from the inherent lovability of his character. Kindly, noble, and chivalrous in the extreme, and though modest and absolutely devoid of mere personal ambition, yet jealous for the interests of his friends and the Society in general, and, in particular, jealous in the interest of everything truly scientific. Anything that looked like ostentation or quackery excited Joule's indignation, particularly when exhibited by those possessing the popular ear. On the other hand, he always noticed with encouragement the efforts of those who were yet unknown, and resented any attempt at the disparagement of their work—as though his own early experience had left him with a fellow-feeling with those who were struggling to get their views taken up.

At this time the Vice-Presidents of the Society were

Mr. Binney, Drs. Schunck and Angus Smith and Mr. Gaskell. Dr. Roscoe (Sir Henry) and Mr. Baxendell were Secretaries. With all these Joule was on terms of intimate friendship. Sir Henry Roscoe's affection for Joule was rather that of a son for his father than of Secretary for President. He used to get Joule, whose life at this time was very retired, to come and dine with him not infrequently, and whenever any foreign or non-resident savant happened to be in Manchester. On such occasions Joule's eminently sociable disposition would appear. He would take a lively part in the discussion of any general topic, and although conservative in his views, would show himself perfectly amiable and tolerant of the views of others. At such, and other times, when his mind unbent, he would chat away and talk intelligent nonsense in discussing a paradox with the greatest pleasure and child-like simplicity.

At this time Joule was in possession of his usual health, and showed no signs of having suffered from his long-continued labours.

In 1870, Joule received that Blue Ribbon of English science—the Copley medal; and in the same year was elected corresponding member of the Academy of Sciences, Institute of France, besides being again Vice-President of the British Association.

The fact that only one Copley medal is awarded each year conveys no idea of the honour which it confers on an Englishman, as it is open to all the world, and it only falls to an Englishman about once in three years. Thus it was awarded Darwin in 1864, Wheatstone in 1868, and Joule in 1870, Silvester in 1880, Cayley in 1882, and Thomson in 1883.

These honours caused the greatest delight in the

Society, and afforded opportunities for observing the complete absence of egotism in Joule's character. He, ignorant of what was coming, was dining with Professor W. C. Williamson, other members of the Society being present, when he first heard that he was to receive the Copley medal. He frankly showed the gratification which he felt in the congratulation and expressions of sympathy from his friends; at the same time it was clear that he accepted the honour as evidence of the recognition of the importance of the truth for which he had so long laboured rather than any special merit of his own.

The numerous short communications which Joule read before the Society between 1869 and 1872 afforded the members opportunities of appreciating the activity and clearness of his mind, and his scientific resource, as well as his facility and skill as an experimentalist. But these communications, relating chiefly to the invention of improvements in physical measuring instruments, undertaken for his own satisfaction and amusement and published mainly to interest the Society, were invariably accepted with appreciation, and afforded little evidence of Joule's more personal character. The Society was, however, early in 1871, to be treated to the opportunity of witnessing Joule under circumstances such as had rarely, if ever, happened to him before. The first occasion was that which called forth from Joule his three communications "On the Alleged Action of Cold in Rendering Iron and Steel Brittle."

The severe winter of 1870, memorable for the Siege of Paris, having attracted attention to the increase of railway fractures, induced numerous comments on the effect of frost on the strength of iron and steel. Some of the members of the Society had taken an interest in the subject, and brought forward experiments which seemed to confirm the effect. This induced Joule to make experiments to disprove at once that pretence, "set up," he says, "to excuse certain Railway Companies" for not having taken extra precautions on account of the extra hardness of the ground, "which is the common-sense explanation of the accidents," but which pretence "although put forth in defiance of all we know of the properties of materials, but also of the experience of everyday life, has yet obtained the credence of many people." His experiments on darning needles showed that at 12° Fahr. the needles were one per cent stronger than at 55° Fahr., and that under the same circumstances the proportion of garden nails at 2° Fahr. which broke was less than the proportion which broke at 40° Fahr. Joule's only motive in taking needles and garden nails, arose from the fact that for his experiment it was necessary to have a considerable number of similar specimens of steel and cast iron, sufficiently small to admit of ready experiments, and the needles and garden nails came readiest to his hand. At the time, however, the idea that engineers would pay attention to experiments on anything so small as darning needles and garden nails, was received with derision by some of those present, and the discussion extended over two more meetings. This gave those present opportunities of seeing Joule under circumstances such as had not happened within the memory of many, if of any, of the members. He was much excited by the opposition, and entered warmly into the discussion, hitting out straight and with spirit, but at the same time with a dignity, courtesy and kindness, which took the sting out of the hard things he said, and was very gratifying to witness.

For Joule's communication on " Examples of the Performance of the Electro-Magnetic Engine," the Society was also indebted to a paper communicated to the Society (not by a member), criticising the conclusions Joule had arrived at in 1845 as to the economic limits of electro-magnetic engines; and although this critical paper was written in ignorance of the subject, the Society owes thanks to the author for having drawn from Joule a dissertation on his earlier work, which, to those who heard it, was a revelation as to what Joule must have been in his earlier days. Joule himself enjoyed the opportunity, and insisted that the paper should be read and the author of the paper should be invited to attend. As may be supposed, there was a strong feeling of indignation amongst the members at the apparent insolence of the attempt, but the discussion was confined to the author and Joule, who not only answered the objections, but, in the most dignified and courteous manner, explained to the author the elementary mistakes he had made, displaying the clearness of his memory as to the details of this complicated research made a quarter of a century before.

Besides the papers already mentioned, Joule made ten other short communications to the Society between 1869 and 1872. Three of these were on his Dip Circle, in which the dip needle was suspended from a balance by spider lines, and the seven others were severally on;—Physical Properties of Bees' Wax;—Photographs of the Sun;—Sunset at Southport;—the Magnetic Storm of February 4, 1872;—the Polarization of Platina Plates by Frictional Electricity;—the Prevalence of Hydrophobia;—and a Mercurial Air Pump; this was the first of the now common displacement pumps.

In addition to the work which these entailed, Joule was at this time engaged preparing his new apparatus for what was to be his last determination of the mechanical equivalent of heat. This research originated in the discrepancy, already mentioned, between the determination by electric currents (782·5), in 1867, and his final determination, in 1849, from the friction of water (772). As this discrepancy ouldc only be accounted for by admitting an error in his thermal experiments, or in the unit of electrical resistance, at the meeting of the British Association in 1869, a committee, consisting of Joule, Sir William Thomson, Professors P. G. Tait, J. C. Maxwell, and Balfour Stewart, was appointed. By this committee Joule was charged with the investigation for verifying his previous results by the direct method, and attaining, if possible, greater accuracy. The preparation for the research took long and was interrupted.

In 1872, Joule was elected President of the British Association for 1873, and accepted the office. He had prepared his address, when, within a few months of the meeting, his health gave way, and the symptoms were so serious that he was advised he would not be able to discharge the duties he had undertaken. The council had recourse to Professor Williamson, who kindly undertook the office at the last moment, and prepared an address, the preamble of which is as follows :—

" Instead of rising to address you on this occasion, I had hoped to sit quietly amongst you and to enjoy the intellectual treat of listening to the words of a man of whom England may well be proud—a man whose life has been spent in reading the book of nature for the purpose of enriching his fellow-men with the knowledge of its truths—a man whose name is known and honoured in every corner of this planet to which a knowledge of science has penetrated—and, let me add, a man whose name will live in the grateful memory of mankind as long as the records

FIRST FAILURE OF HEALTH.

of such noble work are preserved. At the last meeting of the Association I had the pleasure of proposing that Dr. Joule be elected President for the Bradford Meeting, and our council succeeded in overcoming his reluctance and persuading him to accept that office."

"Nobly would Joule have discharged the duties of President had his bodily health been equal to the task; but it became apparent, after a while, that he could not rely upon sufficient strength to justify him in performing the duties of the chair, and in obedience to the orders of his physicians he placed his resignation in the hands of the council about two months ago."

Joule was deeply grateful to Dr. Williamson for so kindly coming to his aid; but his distress at the inconvenience which his illness involved, added much to the severity of his attack. It was nearly three years before he sufficiently recovered to resume either his experimental work or his regular attendance at the meetings of the Society; and it was not till 1877 that he resumed the office of President, which, in the meantime, had been held by Dr. Schunck and Mr. E. W. Binney.

Of the attack, Mr. B. St. J. B. Joule writes, quoting from his diary, 1873, April 6:—" My brother had an attack of bleeding from the nose. Dr. Samuel Crompton arrived (Cliff Point) at 5 a.m. and remained, with two short intervals, till 11 p.m. He inserted a plug." 1874, Feb. 20th (at Southport):—"James was returning from a short walk with me, when bleeding from the nose commenced. I knew that he had given strict injunctions at home that if bleeding occurred again no doctor was to be called in, for he would rather die than submit to the torture he had endured last year. He would not allow a plug to be used on any

M

pretence. I argued the point with my brother. I told him I did not intend to have an inquest held in my house if the worst happened, and a medical man should attend. I recommended Dr. Daniel Elias. At last, I obtained his permission to send for Dr. Elias, but I took care to have a long conversation with him before he saw my brother. I told him I was about to introduce him to a very difficult patient. I informed him of all that had passed, and told him that I was certain he would be obliged to undergo a severe cross-examination, and then took him upstairs. The interview was what I expected. Many questions relating to the case were asked, then came the crucial inquiry, after Dr. Elias had said it might be necessary to introduce a plug, 'but how long would you keep it in?' 'That depends upon circumstances.' 'But would you keep it in for several days?' 'Certainly not, we should change it of course.' The doctor came off with flying colours. It was most providential that this second attack occurred when James was with us, where he was kept quiet and properly attended to." 21st Feb. :—" Dr. Elias called to see James. Arthur (Joule's son) called the same evening." 22nd, 23rd, and 24th :—" Dr. Elias called each day." 25th:—"Called Dr. Elias up at 5 a.m. He called twice this day, and so on up to the 19th of March." Then March 20:—" Dr. Elias called. I accompanied James and Arthur to Manchester. (This was, by far, the worst attack he ever had)." 1874, May 4th :—" James had a return of the bleeding. He was hesitating which of his medical friends, members of the Literary and Philosophical Society, he should call in, but I spoke strongly in favour of Dr. Dixon Mann, and I received permission to call Dr. Mann, of which I at once availed myself by calling upon him."

Note, 1891 :—"I have never had occasion to regret my advice, Dr. Mann soon obtained my brother's confidence, and proved a most attentive and judicious friend as well as medical adviser. I think I may fairly say that I had the privilege of saving my brother's life for a quarter of a century.

In August and September, 1873, Joule spent six weeks with his family at Douglas; in 1874 he occupied himself in making slight improvements in the mercurial displacement pump, which he had invented before his health gave way, and in 1875, he read a short paper "On a Glue Battery."

It was about this time that he resumed his work on the "Verification of the Mechanical Equivalent of Heat," commenced for the British Association. But owing to unfortunate investments, his income having become much diminished, he found himself unable to undertake the very considerable expense the investigation involved over and above the small grant originally made by the Association. This becoming known, the Royal Society made him a grant of £200 out of the fund of £4,000 placed at their disposal by the Government. With this assistance, Joule during the years 1876—7 perfected the apparatus and made the experiments, devoting himself entirely to the work. The results were published in a paper "New Determination of the Mechanical Equivalent of Heat," which was read January 24th, 1878, and printed in the *Phil. Trans.*, 1878, Part II.

In these experiments the old thermometers were again used. Before using them, however, the values of the graduations were again determined with the utmost care, and with every precaution experience could suggest, the result being that although the freezing point had risen

1·84 divisions or 0°·14 Fah., the value of the divisions was unaltered—the new determination giving 0°·077,223 Fah. as the value of a division, while the old was 0°·077,214.

The apparatus for agitating the water and measuring the power used in these experiments was entirely new, and although in designing it Joule had availed himself to the utmost of his previous experience, the new apparatus differed essentially from the old.

The most important difference was in the method of applying and measuring the work. In his experiments of 1849 Joule had used descending weights, like clock weights, to turn the paddle, the power so employed being determined as the product of the weight multiplied by the fall. The disadvantage of this method arose from the necessity of winding up the weight some 20 times for each experiment, and the time thus taken up increasing the time and radiation effects during the experiments. To obviate this it was necessary to realize the principle that with a steady motion of the paddle, the turning moment on the paddle would be exactly equal to the resistance necessary to hold the vessel. This principle Joule realized for himself, and delicately poising his vessel on a pivot, by passing silk threads round a pulley on the pivots, he determined the resistance of the vessel, and then by turning the handle of the paddle so as just to balance the fixed resistance, and counting the revolutions he measured the work done. This invention was precisely the same as had previously been made by Hirn, and used by him in verifying Joule's Equivalent in 1865, and was shortly after almost simultaneously employed by Froude and others for measuring the work of steam engines. With respect to it, Joule says: "The plan I adopted was, in regard to the measurement of work, similar to that used by Hirn, who has laboured so earnestly and successfully on this subject. He has described it as follows "—with an extract of Hirn's paper.

In this research Joule made five series of experiments at different speeds, each series involving from 6 to 21 separate determinations. The agreement between the results so obtained was very close, and on carefully reducing these to a mean, the result arrived at was 772·55, which differed from his final determination in 1849, 772·692, by less than two inches of fall, or one part in five thousand, thus verifying his final result 772 to within a thousandth part, and establishing it as being, as yet, the most accurate, as it is the most difficult, of all the accurate and difficult determinations which have gone to define the dynamical theory of matter. Considering that the powers of observation and contrivance required to obtain this accuracy are such that none of the many eminent experimental physicists who have worked since 1850 have rivalled Joule in his work, the repetition of his experiment of 1849 in 1878, after an interval of 29 years, and at the age of sixty, is an unprecedented performance, and shows, more emphatically than was otherwise possible, the special powers as an experimentalist with which he was endowed, and of which advancing years had, as yet, in no way deprived him.

This was the last experimental research of any magnitude that Joule undertook. In 1877, he again changed his residence; having purchased No. 12, Wardle Road, Sale, he removed into it, and resided there till his death, using an out-building and the cellars of his house for the purposes of keeping his apparatus and making experiments. He paid two visits this year to Mr. Binney, at Douglas, and was there again in 1879.

He persuaded Mr. Binney to sit to Mr. W. H. Johnson for the admirable portrait which Joule presented to the Society.

In June, 1878, he received a letter from the Prime Minister, Lord Beaconsfield, announcing to him that her Majesty the Queen had been pleased to grant him a pension of £200 per annum, which recognition of his labours by his country was a source of much gratification to him as well as being a great convenience.

He was receiving Scientific Honours with increasing frequency. He was elected in 1873 Corresponding Member of the Academies of Copenhagen and of Bologna; American Academy, Boston, 1874; B.H.D. University, Leyden (300th Anniversary) 1875; Hon. Mem. Literary and Antiquarian Society, Perth, 1876; "Sociétiè Français de Physique," Paris, 1878; Foreign Corresponding Member, Royal Academy of Science, Turin, and Member (Hon.) of the Academy of Sciences, New York, 1879.

In 1880 he was elected a Member of the First Court of the Victoria University, Manchester, the meetings of which he regularly attended. He also received the Royal Albert Medal from the hand of the Prince of Wales on behalf of the Society of Arts.

In 1881 the Council of the Physical Society of London requested Dr. Joule to allow them to publish a collected edition of his works. To this honourable request Joule readily agreed, and undertook the personal labour of getting them together, and editing the collection.

The first volume of Joule's "Scientific Papers," containing all papers which appeared in his own name, was published in 1884. Then, for the first time, the scientific world had the opportunity of becoming acquainted, not only with the full scope and completeness of Joule's research, but what was more important, with the extent to which Joule had himself anticipated those who followed him in generalizing from his own results. Up to that time Joule's work was mostly known at second hand, or by reference to the paper containing the final determination of the Mechanical Equivalent of Heat, in the *Phil. Trans.*, 1850. But few authors have referred to any of the series of Joule's early discoveries by which he was led to the recognition of the Conservation of Energy almost before he had determined the Equivalent. These discoveries excited no remark at the time they were made, and while Joule was pursuing the equivalent others had taken up his previous work, and these discoveries, including that of the Universal Conservation of Energy, had been swallowed up in the rising tide of science, from which they were only rescued by the appearance of this volume. Until this volume appeared it is improbable that any one of the many eminent physicists then living had any knowledge of Joule's lecture at the St. Ann's Church Reading Room, published in the *Courier* in 1847, containing the only full expression of his views on the indestructibility of energy.

The second volume of Joule's "Scientific Papers" appeared in 1887. This contained his joint papers with the Rev. Dr. Scoresby, Sir Lyon Playfair, and Sir William Thomson.

Joule not only collected his papers and edited these volumes, adding numerous notes containing references to subsequent researches on the various subjects, both his own and others, but also added accounts of several of his researches not previously published. The most important of these, is, perhaps, his "Account of Experiments on

Magnets, begun in 1864 at Old Trafford (written in 1882-83)." In this research Joule subjected the magnetic permanence of numerous steel, wrought and cast iron magnets, and two loadstones, to observations extending over 18 years, besides making numerous experiments on the effect of temperature on the intensity of the magnets.

This work of editing fully explains why Joule ceased his small scale experimental work and his contributions to the Society after 1879, when he read his last paper on " A Method of Checking the Oscillations of a Telescope."

Although he never fully recovered his health, he was still active. He was President in 1878, and again in 1879, for the last time, as he declined to take the office in 1882.

During this time he paid frequent visits with his son to his brother in Rothsay. In 1882 he was for some time in London sitting to Collier for his portrait, which was subsequently presented to the Royal Society by the members of the Council. In 1883 he and his son spent some time in North Devon.

The death of his oldest friend and colleague in the Society, Mr. E. W. Binney, which occurred in December, 1881, was a great shock to him; and the Society was never quite the same to Joule afterwards. They were elected members of the Society the same day, January 25, 1842, had been co-Secretaries for several years, Vice-Presidents from 1851 and 1852, and each had held the office of President for 10 years. Though essentially different in temperament, these men had a common ground of sympathy in their love of all that was real in knowledge, and their dislike of all pretence, and there was real love between them. Joule, however, had still many—most of his—friends among the members, and maintained his

FAILING HEALTH—DEATH.

interest in the Society ; even after the death of Dr. Angus Smith, in 1885, and the absence of Sir Henry Roscoe and Dr. Schunck, although the *personnel* of the Council became greatly altered, Joule still attended the meetings whenever he could. In November, 1885, he dined with Dr. Schuster on the occasion of a visit from Lord Rayleigh. On March 9, 1886, he attended both the Council and the Society for the last time, but his signature appears in the library book as late as October 9. It was then evident that his strength was giving way, but as his mind was perfectly clear this did not at first excite alarm.

In 1886 he was again solicited to accept the office of President of the British Association at the meeting in Manchester in the following year, but at that time his health was such that he felt it was out of the question. He now spent much of his time with his daughter at Seaforth, and his brother induced him to be there during the meeting in 1887, lest the excitement resulting from friends calling should be too much for him. While there, during the meeting, he made the following characteristic remark to his brother :—

" They" (Arthur and Alice) "have been telling me that Roscoe has been saying some very good things about me. I wish they would let me alone. I believe I have done two or three little things, but nothing to make a fuss about."

From this time the deterioration of his mental powers commenced, and he scarcely left his house or saw anyone but his family. After a protracted but painless illness, during the last days of which he was nearly unconscious, James Prescott Joule, Discoverer of the Universal Conservation of Energy, passed away in the midst of his family, at his residence, 12, Wardle Road, Sale, on the evening of Friday, October 11th, 1889.

Joule was buried in the Western Cemetery on Wednesday, October 16th. The funeral was strictly private, no public announcement having been made as to when or where it was to take place. The Society was represented by its officers and other members, but besides these only the family were present. Occurring late on Friday evening the first intimation of Joule's death which was received, outside his household, was on the Monday following, when obituary notices appeared in the papers. On the same day the President of the Society was privately informed of the arrangements which had been made for the funeral. These he communicated to the Council and Society at their meetings, which took place in the ordinary course on Tuesday, October 15, when resolutions of condolence with the family were passed, and deputations appointed to attend the funeral.

At the same meetings resolutions were passed authorising the preparation of a full memoir of Joule's Life and Works, and appointing a committee to inaugurate the necessary steps to secure a fitting memorial of Joule in Manchester.

The committee had to consider the question whether the memorial should be in the Society, or whether an effort should be made to secure a public memorial in Manchester. The latter course commended itself, and steps were immediately taken to secure signatures to a requisition to the Mayor to call a Town's Meeting. At first there was considerable fear lest Joule's extremely retired life had left his fellow-citizens ignorant of the honour which accrued to the city and district from his life-long association. These fears were, however, found to be groundless.

The Mayor, Mr. Alderman Mark, took up the matter cordially, and over three hundred signatures, including those of all the leading residents, were appended to the requisition. The meeting was held on November 25th, 1889, in the Mayor's Parlour, and was largely and influentially attended, and it was unanimously resolved to have a memorial of Joule, in the form of a white marble statue, to be a companion to Chantrey's statue of Dalton. A committee having been appointed to raise funds, the money was readily subscribed, the list of subscribers containing three hundred names; and the execution of the statue was entrusted to Mr. Alfred Gilbert, A.R.A., whose model was approved in the summer of 1891.

Immediately after Joule's death it was felt that it would be only fitting that a national memorial should be placed in Westminster Abbey, and the Council of the Royal Society having taken the initiative, the necessary authority has been obtained to erect a memorial tablet immediately adjacent to the medallion in memory of Darwin. At the same time the Council of the Royal Society has raised a fund for the establishment of a memorial of an international character commemorative of the life and work of Joule, to have for its object the encouragement of research in Physical Science.

There are two known portraits in oil of Joule. One by G. Patten, 1864, in the possession of the Society, from which the engraving prefixed to this Memoir is taken; the other by Collier, 1882, in the possession of the Royal Society. There is also the excellent engraving by Jeens, published in 'Nature,' in 1882, and again in 'Joule's Scientific Papers.' There are also several very good photographs taken in later life, particularly one taken by Lady

Roscoe. Mr. A. Joule also has in his possession a bust of his father. With the exception of the portrait by Patten, all these represent Joule in later life, and after he had grown a beard. Patten's portrait, which represents him at the age of 45 and in full vigour, has so far been little known, except within the Society, and it is chiefly for this reason that the Council have taken the opportunity afforded of publishing the engraving as a frontispiece to this Memoir.

APPENDIX TO PAGE 18.

DEVELOPMENT OF KNOWLEDGE OF HEAT FROM 1650.—*Hooke's Vibratory Theory of Heat and Light.—Heat as a measurable Fluid.—Heat as an indestructible Fluid.—"Caloric"; invented by Lavoisier.—Extracts from "Micrographia," by Hooke.—Extracts from "Traité Elémentaire de Chimie," by Lavoisier.*

The following extracts from the *Micrographia* show that, already in 1660, Robert Hooke had carried the mechanical theory of heat out of the region of speculation, and adduced experimental proofs that heat and light have their origin in the invisible vibratory motions of matter, and are communicated through space by the vibratory motions of the ether; and had given such definition to his theory that he was thereby enabled to give the now accepted explanations of fluidity and combustion, and to point to the deflection of the wave fronts ("pulses in orbem") when waves pass obliquely from one medium into another in which they move with a different velocity.

The papers from which these extracts are taken were read before the Royal Society between the years 1660 and 1664, when the *"Micrographia"* was published, and it is on record (" Posthumous Works of Robert Hooke," by Richard Wallis, 1705, p. 8) that these papers made Hooke much respected by the Society. It appears that, for some time after these views had been expressed, heat was very generally regarded by the advanced school of natural philosophers as the result of external motions of matter. Robert Boyle, in the preface to his work "An Experimental History of Cold," published in 1664, says:—"*The subject I have chosen is very noble and important; for since Heat has so general an interest in the Production of Nature's Phænomena, that (Motion excepted, of which it is a kind) there is scarce anything in Nature whose efficiency is so great and so diffused, it seems not likely that its antagonist,* Cold, *should be a despicable Quality.*" Huygens also adopted views similar to those of Hooke; and these views seemed to be in a fair way of general acceptance until the brilliancy of Newton's mathematical discoveries diverted thought, and established a school which accepted the corpuscular theory as proved, rejecting the vibratory theory as unorthodox—not having the sanction of their illustrious founder. Hooke's concrete idea of heat as motion was thus dropped, and the vague ideas of an igneous fluid revived; but these were not allowed to remain a mere matter of speculation. The not indefinite hypothesis of heat as a material under the name "phlogiston," was subjected to the test of explaining the then known chemical actions during the first half of the eighteenth century, and held no inconsiderable place in professorial lectures as late as those of Black. These tests introduced the system of weighing and measuring in chemistry, and amongst others of measuring heat by the degree of temperature it would impart to a definite quantity of water,—which was certainly the first and most important step in the science of heat. Then followed, as a direct consequence, Black's discovery, about 1762, of the disappearance of definite quantities of heat during the changes of state, fusion and evaporation, and the respective reappearance of the same quantities of heat during the reverse changes, freezing and condensation. Black called this heat, which had disappeared to reappear when the action was reversed, "latent heat," thereby implying that the heat which had disappeared was still present in the matter. This view which, though now known not to be necessarily true, received general acceptance, and seemed to afford what amounted to an indisputable proof of the indestructible or material character of heat; and, being advertised by the inventions of Watt, met with general recognition. The nature of heat, from being a subject of speculation, was thus placed on what seemed to be a sound foundation as an elastic fluid, which, besides being indestructible and uncreateable, was definitely measurable. From this time the science of heat, both as to its agency in the various chemical actions and as to the laws of its communication by radiation and conduction, made steady progress. To emphasize the greater definition and certainty of the new hypothesis, Lavoisier and the French Academicians, in 1787, invented the name "Calorique," by which to distinguish heat as a quantity from the temperature and sensations with which it is associated, and so introduced great simplification in the science. This simple definition of heat, however consistent it might be with its real mechanical nature, was in the main accepted literally, and caused heat to be regarded as material, and this view seems to have found confirmation during the rapid advances of experimental science, mostly chemistry, which were taking place. The assumption that whatever heat reappeared or disappeared had been or had become latent was so reasonable whenever (as in most cases) the phenomenon was the result of some physical or chemical changes, that the few and less interesting cases in which heat appeared as the result of mechanical action, such as rubbing, which produced no sensible changes, were met by metaphysical assumptions which were allowed to pass, and gradually acquired a place in the teaching of the schools as sanctioned by authority. These were accepted, however, only by those who did not study the phenomena, which did not then command much attention. That the steady stream of heat which may be poured out of a body continuously by the simple process of rubbing it should be the result of occult changes in the body could be held by no true philosopher who had the phenomenon brought forcibly before him. Thus Rumford, in 1791, followed by Davy and Young, revived the dynamical theory, and again showed heat to have a mechanical origin. But although they spoke from the Royal Society and Royal Institution, and although nobody seems to have doubted the truth or force of what they said, they could get no considerable following.

The definite hypothesis of caloric had been found so fitting, had explained so much, and had been such a great simplification, that it was not lightly to be abandoned for a view, which, so far as it was then carried, promised to throw everything again into chaos. Heat might be mechanical, but it was certain that it was measurable, and, for the most part, had a permanent existence; in these respects, and many others, the hypothesis of caloric had been found a safe guide, and such guidance was more important than mere views as to the nature of heat. Such views, however incontestable might be the evidence adduced, must be open to doubt so long as they appeared contrary to what was known to be true. Those who advocated the mechanical theory had not shown how, in any definite measure, mechanical effect could correspond with heat, while experience had established as an axiom, that perpetual motion was impossible, and that mechanical effect was destructible, how then could heat, which was measurable and continued to exist, be mechanical force? This question had to be answered before the caloric hypothesis could be abandoned, if, indeed, it can be said to have been abandoned, because the name, and certain assumptions tacked on are dropped, while all that was clearly defined in the hypothesis, having been found true and in accordance with the mechanical theory, is retained. That "caloric" was not intended to convey an idea of the materiality of heat is clearly shown in the extract (which follows those from Hooke) from the "Traité Elémentaire de Chimie," by Lavoisier, who invented the name, and that it did not necessarily convey this idea is shown by the number of philosophers, who, though for long unconvinced of the conclusions drawn from the mechanical theory, still disclaimed any belief as to the materiality of heat.

Extract from "Micrographia," by Robert Hooke, p. 12:—

"First what is the *cause* of *fluidness?* And this I conceive to be nothing else but a certain *pulse* or *shake* of heat; for heat being nothing else but a very *brisk* and *vehement agitation* of the parts of a body (as I have elsewhere made probable) the parts of a body are thereby made to loose from one another that they easily *move any way* and become fluid. That I may explain this by a gross Similitude. Let us suppose a dish of sand set upon some body that is very much agitated and shaken by some *quick* and *strong vibrating motion,* as on a *milstone* turned round upon the understone, very violently, whilst it is empty, or on a very stiff *drum*-head which is vehemently or nimbly beaten with

Drumsticks. By this means the sand in the dish which before lay like a *dull* and inactive body, becomes a perfect *fluid;* and you can no sooner make a *hole* in it with your finger, but it is immediately *filled up again* and the upper surface of it *levell'd*. Nor can you bury a *light body*, as a piece of cork, under it, but it presently *emerges* and *swims* as it were upon the top; nor can you lay a heavier on the top of it, as a piece of lead, but it is immediately buried in sand and (as it were) sinks to the bottom. Nor can you make a hole in the side of the Dish but the sand will *run out* to a *level*, not an *obvious property* of a fluid body, as such, but this does imitate; and all this merely caused by the vehement *agitation* of the containing vessel; for by this means *each* sand becomes to have a *vibrating* or *dancing* motion so as no other heavier body can *rest* upon it unless sustained by some other on either side; nor will it suffer any body to be beneath it unless it be a *heavier* than itself. Another instance of the strange loosening nature of a violent jarring motion or a strong and nimble vibrative one, we have from a piece of iron grated on very strongly with a file; for if into that a pin be *screwed* so firm and hard that, though it has a convenient head on it, yet it can by no means be *unscrewed* by the fingers; if, I say, you attempt to unscrew this whilst *grated on by the file* it will be found to undo and turn very easily. The first of these examples manifests, how a body actually *divided* into small parts, becomes a fluid. And the latter manifests by what means the agitation of heat so easily loosens and unties the parts of *solid* and *firm* bodies. Nor need we suppose heat to be anything else, besides such a motion; for supposing we could mechanically produce such a one quick and strong enough, we need not spend fuel to melt a body. Now that I do not speak of this altogether groundless, I must refer the Reader to the Observations I have made on the shining sparks of Steel, for there he shall find that *the same* effects are produced upon small chips or parcels of Steel by the *flame* and by *a quick and violent motion;* and if the body of *Steel* may thus be melted (as I there show it may) I think we have little reason to doubt that almost *any other* may not also. Every Smith can inform one how quick both his *File* and the *Iron* grow *hot* with *filing*, and if you *rub* almost any two *hard* bodies together they will do the same, and we know that a sufficient degree of heat causes *fluidity* in some bodies much sooner and in others later; that is the parts of some are so *loose* from one another and are so *unapt to cohere*, and so *minute* and *little*, that a very *small* degree of agitation keeps them always in a *state of fluidity*. Of this kind, I suppose the *Æther*, that is the *medium* or *fluid* body in which all other bodies do as it were swim and move, and particularly the air."—(*ib.* p. 16.) "Now that the *parts* of all *bodies*, though never so *solid*, do yet *vibrate*, I think we need go no further for proof than that *all* bodies have some *degree* of *heat* in them, and that there has not been yet found anything *perfectly cold*. Nor can I believe indeed that there is any such thing in Nature as a body, whose particles are at *rest* or *lazy* and *unactive* in the great *Theatre* of the *World*, it being quite contrary to the grand *Œconomy* of the Universe."

Extract from "*Micrographia*," *p. 44*:—

"It is a very common Experiment, by striking with a Flint against a Steel, to make certain fiery and shining sparks to fly out from between those two compressing Bodies.

"About eight years since, upon casually reading the explication of this odd *Phenomenon*, by the most ingenious Des Cartes, I had a great desire to be satisfied what that Substance was that gave such a shining and bright light; and to that end I spread a sheet of white paper, and on it, observing the place where several of these sparks seemed to vanish, I found certain very small, black, but glistening spots of a movable substance, each of which, examining with my *Microscope*, I found to be a small, round *Globule*, some of which, as they looked pretty small, so did they from their surface yield a very bright and strong reflection on that side which was next the Light, and each looked almost like a pretty bright iron ball, whose Surface was pretty regular. In this I could perceive the Image of the Window pretty well, or of a Stick which I moved up and down between the Light and it. Others I found which were, as to the bulk of the ball, pretty regularly round, but the Surface of them, as it was not very smooth, but rough, and more irregular, so was the reflection from it more faint and confused." (*Ib.* p. 45.) "He that shall diligently examine the *Phænomena* of this Experiment will, I doubt not, find cause to believe, that the reason I have heretofore given of it is the true and genuine cause of it, namely, That *the Spark appearing so bright in the falling, is nothing else but a small piece of the Steel or Flint, but most commonly of the Steel, which, by the violence of the Stroke, is, at the same time, severed and beatt red hot, and that sometimes to such a degree as to make it melt together into a small Globule of Steel; and sometimes also is that heat so very intense, as further to melt it and vitrifie it; but many times the heat is so gentle as to be able to make the sliver only red hot, which, notwithstanding falling upon the tinder* (that is only a very curious small coal made of the small threads of Linnen burnt to coals and char'd) *it easily sets it on fire.*" Nor will any part of this Hypothesis seem strange to him that considers—First, that either hammering, or filing, or otherwise violently rubbing of steel will presently make it so hot as to be able to burn one's fingers. Next, that the whole force of the stroke is exerted upon that small part where the Flint and Steel first touch: for the bodies being each of them so very hard, the puls cannot be far communicated, that is, the parts of each can yield but very little, and therefore the violence of the concussion will be *exerted* on that piece of steel which is cut off by the flint."

Extract from "Micrographia," p. 54:—

" And first for Light, it seems very manifest, that there is no luminous body but has the parts of it in motion more or less.

" First, that all kinds of *fiery burning Bodies* have their parts in motion, I think will be very easily granted me. That the *spark* struck from a Flint and Steel is in a rapid agitation, I have elsewhere made probable. And that the parts of *rotten wood, rotten fish,* and the like, are also in motion, I think, will as easily be conceded by those who consider that those parts never begin to shine till the bodies be in a state of putrefaction; and that is now generally granted by all, to be caused by the motion of the parts of putrifying bodies. That the *Bononian stone* shines no longer than it is either warmed by the sunbeams, or by the flame of a fire, or of a candle, is the general report of those that write of it, and of others that have seen it, and that heat argues a motion of the internal parts, is (as I said before) generally granted.

"But there is one instance more, which was first shown to the *Royal Society* by *Mr. Clayton*, a worthy member thereof, which does make this assertion more evident than all the rest: and that is, that a *Diamond* being *rubbed*, *struck*, or *heated* in the dark, shines for a pretty while after, so long as that motion, which is imparted by any of those agents, remains (in the same manner as a glass rubbed, struck, or (by a means which I shall elsewhere mention) heated, yields a sound which lasts as long as the *vibrating* motion of that *sonorous* body), several experiments made on which Stone, are since published in a 'Discourse of Colours,' by the truly honourable *Mr. Boyle*." (*ib.* p. 55). "It would be too long, I say, here to insert the discursive progress by which I inquired after the proprieties of the Motion of Light; and therefore I shall only add the result.

"And, first, I found it ought to be exceedingly *quick*, such as those motions of *fermentation* and *putrefaction*, whereby, certainly, the parts are exceedingly nimbly and violently moved; and that, because we find those motions are able more minutely to shatter and divide the body than the most violent heats or *menstruums* we yet know. And that fire is nothing else but such a *dissolution* of the Burning Body, made by the most *universal menstruum* of all *sulphurous bodies*, namely, the Air, we shall, in another place of this Tractate, endeavour to make probable. And that, in all extremely hot, shining bodies, there is a very quick motion that causes Light, as well as a more robust that causes Heat, may be argued from the celerity wherewith the bodyes are dissolved.

"Next, it must be a *vibrative motion*. And for this the newly mention'd Diamond affords us a good argument; since if the motion of the parts did not return, the Diamond must after many rubbings decay and be wasted ; but we have no reason to suspect the latter, especially if we consider the exceeding difficulty that is found in cutting or wearing away a Diamond. And a circular motion of the parts is much more improbable, since, if that were granted, and they be suppos'd irregular and angular parts, I see not how the parts of the Diamond should hold so firmly together, or remain in the same sensible dimensions which yet they do.

"Next, if they be *Globular*, and mov'd only with a turbinated motion, I know not any cause that can impress that motion upon the *pellucid medium*, which yet is done. Thirdly, any other *irregular* motion of the parts one amongst another, must necessarily make the body of a fluid consistence, from which it is far enough. It must therefore be a *vibrating* motion.

"And thirdly, that it is a very *short vibrating motion*, I think the instances drawn from the shining of Diamonds will also make probable. For a Diamond being the hardest body we yet know in the world, and consequently the least apt to yield or bend, must consequently also have its *vibrations* exceeding short, and these, I think, are the three principal proprieties of a motion requisite to produce the effect called light in the object.

"The next thing we are to consider is the way or manner of the *trajection* of this motion through the interpos'd pellucid body to the eye. And here it will be easily granted,

"First, That it must be a body *susceptible* and *impartible* of this motion that will deserve the name of a Transparent. And next, that the parts of such a body must be *Homogeneous*, or of the same kind. Thirdly, that the constitution and motion of the parts must be such that the appulse of the luminous body may be communicated or propagated through it to the greatest imaginable distance in the least imaginable time; though I see no reason to affirm that it must be in an instant, for I know not any one experiment or observation that does prove it. And, whereas, it may be objected, That we see the Sun risen at the very instant when it is above the sensible Horizon, and that we see a Star hidden by the body of the Moon at the same instant when the Star, the Moon, and our Eye are all in the same line; and the like Observations, or rather suppositions, may be urg'd. I have this to answer, That I can as easily deny as they affirm; for I would fain know by what means any one can be assured any more of the Affirmative than I of the Negative. If, indeed, the propagation were very slow, 'tis possible something might be discovered by Eclypses of the Moon; but though we should grant the progress of the light from the Earth to the Moon, and from the Moon back to the Earth again, to be full two minutes in performing, I know not any possible means to discover it; nay, there may be some instances, perhaps, of Horizontal Eclypses that may seem very much to favour this supposition of the slower progression of Light than most imagine. And the like may be said of the Eclypses of the Sun, &c. But of this only, by the by. Fourthly, that the motion is propagated every way through an *Homogeneous medium* by *direct* or *straight* lines extended every way like rays from the centre of a sphere. Fifthly, in an *Homogeneous medium* this motion is propagated every way with *equal velocity*, whence necessarily every *pulse* or *vibration* of the luminous body will generate a sphere, which will continually increase and grow bigger, just after the same manner (though indefinitely swifter) as the waves or rings on the surface of the water do swell into bigger and bigger circles about a point of it where, by the sinking of a stone, the motion was begun; whence it necessarily follows that all the parts of these spheres undulated through an *Homogeneous medium* cut the rays in right angles.

"But because all transparent *mediums* are not *Homogeneous* to one another, therefore we will next examine, how this pulse or motion will be propagated through differingly transparent *mediums*. And here, according to the most acute and excellent Philosopher *Des Cartes*, I suppose the sign of the angle of inclination in the first *medium* to be the sign of refraction in the second, as the density of the first to the density of the second. By density I mean not the density in respect of gravity (with which the refractions or transparency of *mediums* hold no proportion) but in respect only to the *trajection* of the Rays of light, in which respect they only differ in this, that the one propagates the pulse more easily and weakly, the other more slowly, but more strongly."

Extract from " Traité Elémentaire de Chimie," Lavoisier, p. 4.:—

"On en peut dire autant de tous les corps de la Nature ; ils sont ou solides, ou liquides, ou dans l'état élastique et aériforme, suivant le rapport qui existe entre la force attractive de leurs molécules et la force répulsive de la chaleur, ou, ce qui revient au même suivant le degré de chaleur auquel ils sont exposés.

"Il est difficile de concevoir ces phénomènes sans admettre qu'ils sont l'effet d'une substance réelle et matérielle, d'un fluide très-subtil qui s'insinue à travers les molécules de tous les corps et qui les écarte, et en supposant même que l'existence de ce fluide fût une hypothèse, on verra dans la suite qu'elle explique d'une maniere très-heureuse les phénomènes de la Nature.

"Cette substance, quelle qu'elle soit, etant la cause de la chaleur ; ou en d'autres termes la sensation que nous appelons chaleur, étant l'effet de l'accumulation de cette substance on ne peut pas, dans un langage rigoureux la désigner par le nom de chaleur ; parce que la même dénomination, ne peut pas exprimer la cause et l'effet. C'est ce qui m'avoit déterminé, dans le Mémoire que j'ai publié en 1777 (*Recueil de l'Académie*, page 420) à la désigner sous le nom de fluide igné et de matière de la chaleur. Depuis, dans le travail que nous avons fait en commun, M. de Morveau, M. Berthollet, M. de Fourcroy et moi sur la réforme du langage chimique, nous avons cru devoir bannir ces periphrases qui alongent le discours, qui le rendent plus traînant, moins précis, moins clair, et qui souvent même ne comportent pas des idées suffisamment justes. Nous avons en conséquence désigné la cause de la chaleur, le fluide éminemment élastique qui la produit, par le nom *calorique*. Indépendamment de ce que cette expression remplit notre objet dans le système que nous avons adopté, elle a encore un autre avantage c'est de pouvoir s'adapter à toutes sortes d'opinions, puisque rigoureusement parlant, nous ne sommes pas même obligés de supposer que le calorique soit une matiére réelle ; il suffit, comme on le sentira mieux par la lecture de ce qui va suivre, que ce soit une cause répulsive quelconque qui écarte les molécules de la matière et on peut ainsi en envisager les effets d'une maniere abstraite et mathematique."

NOTE *A* TO PAGE 88.

REFERENCE TO THE VIEWS OF ROGET AND FARADAY.—
"*Force*," *destructible and* "*Force*," *indestructible.—Extract
from Faraday's* "*Experimental Researches in Electricity.*"

The view that "*force*" (as Joule both conceived and defined force) is indestructible, to which view Joule was led in 1843, by the discovery of its conversion into an equivalent of heat constituted the final and crowning step in the discovery of the law of the conservation of energy. It was also the step which met with the greatest obstruction from preconceived opinions and the experience associated with what was vaguely known as "force." Joule had generalised "force" to include whatever is now called energy; but according to the then general use "force" included only that portion of energy, which, in virtue of its condition, could be converted into available effect, which portion of energy, after the reconversion by friction of the effect into heat, having lost its condition of availability, ceased to be "force." Thus, according to the general meaning attached to "force," it was universally and rightly held to be destructible, and the indestructibility of "force" only became true when the conception of "force" had been extended as in Joule's mind.

Considering this, there is something very remarkable in the passage from Joule's paper of 1844 :—" Believing, as I do, that the power of destroying things belongs to the Creator alone, I entirely coincide with Roget and Faraday in the opinion that any theory, which when carried out, demands the annihilation of force is necessarily erroneous." In this Joule not only concedes to Roget and Faraday priority in holding the view that force, as he conceives it, is indestructible, but concedes it in such a manner as to imply that their views were so well known that references to their works were unnecessary. He apparently brings their views forward as affording authority for his own. In his first statement of his views as to the indestructibility of "force"—written ten months previously—"being satisfied that the grand agents of nature are, by the Creator's fiat, indestructible, and that whenever mechanical force is expended an exact equivalent of heat is always obtained," he cites no human authority whatever, and as it cannot be doubted that Joule would then gladly have adduced such authority, it must be inferred that he had in the meantime become aware of some expressions by both Faraday and Roget, which were not previously known to him, and which he conceived to be evidence of views in accordance with his own. What these expressions were it is of course impossible to say, but it is important to notice, that previous to June, 1844, and for at least some years after, there appears to be nothing written by Roget or Faraday that contains even a suggestion that they conceived or used "force" in any but the then usual sense as available energy, or that from which available powers could be obtained; nor does it appear that they in any way fell into the error of conceiving this "force" to be indestructible, but, on the contrary, they both emphatically and explicitly express their conviction (of the truth) namely, that wherever mechanical effect is produced "force" is expended, which nothing will revive but an act of creation. These views are expressed clearly in paragraphs 2071 and 2073 of Faraday's "Experimental Researches," to which also Faraday has added a note containing an extract from Roget's work in which the same views are expressed; and there can be little doubt (as was pointed out to the Author, by Mr. F. J. Faraday) that it was these paragraphs to which Joule referred, as they appeared in the *Phil. Trans.*, 1840, and in the reprint, Vol. II., 1844. How Joule managed to misinterpret them is difficult to conceive, except that with his mind full of his own view of "force," he seized the view expressed as to the impossibility of creating "force" as implying views as to the impossibility of its annihilation. That Joule, with his clear insight as to the indestructibility of "force," as he conceived it, was yet blind to the truth of the universal destructibility of "force" as then generally understood, is shown by the many arguments against Carnot's expression of the law of this destructibility in which the passage occurs. And misunderstanding the truth in Carnot's views, it seems possible that he may have interpreted views which, as expressed, were in strict accordance with Carnot's, as being directly adverse, and in favour of his own.

Extract from Faraday's "Experimental Researches in Electricity," 2071 and 2073, with note to 2071:—

"The contact theory assumes, in fact, that a force which is able to overcome a powerful resistance, as for instance that of conductors good or bad, through which the current passes, and that again of the electrolytic action where bodies are decomposed by it, can arise out of nothing; that without any change in the acting matter or the consumption of any generating force, a current can be produced which shall go on for ever against a constant resistance, or only be stopped as in the voltaic trough by the ruins which its exertion has heaped up in its own course. This would indeed be *a creation of power*, and is like no other force in nature. We have many processes by which the form of the power may be so changed that an apparent conversion of the one into the other takes place. So we can change chemical force into electric current, or current into chemical force. The beautiful experiments of Seebeck and Peltier show the convertibility of heat and electricity; and others by Oersted and myself show the convertibility of electricity and magnetism. But in no cases, not even those of the Gymnotus and Torpedo (1790), is there a pure creation of force, a production of power without a corresponding exhaustion of something to supply it.²"

"2073.—Were it otherwise than it is, and were the contact theory true, then, as it appears to me, the equality of cause and effect must be denied (2069). Then would

---

" ² (*Note March 29th, 1840*).—I regret that I was not before aware of most important evidence for this philosophical argument, consisting of the opinion of Dr. Roget, given in his Treatise on Galvanism in the Library of Useful Knowledge, the date of which is January, 1829. Dr. Roget is, upon the facts of science, a supporter of the chemical theory of excitation, but the striking passage I desire now to refer to is the following: At § 113 of the article Galvanism, speaking of the voltaic theory of contact, he says—'Were any further reasoning necessary to overthrow it, a forcible argument might be drawn from the following consideration: If there could exist a power having the property ascribed to it by this hypothesis, namely, that of giving a continual impulse to a fluid in a constant direction without being exhausted by its own action it would differ essentially from all other known powers in Nature. All the powers and sources of motion with the operation of which we are acquainted when producing their peculiar effects, are expended in the same proportion as those effects are produced, and hence arises the impossibility of obtaining, through this agency, perpetual motion. But the electromotive force ascribed by Volta to the metals is a force which, as long as a free course is allowed to the electricity it sets in motion, is never expended, and continues to be excited with undiminished power in the production of a never-ceasing effect. Against the truth of such a supposition, the probabilities are all but infinite.'—*Roget.*"

the perpetual motion also be true; and it would not be at all difficult, upon the first given case of an electric current by contact alone, to produce an electro-magnetic arrangement, which, as to its principle, would go on producing mechanical effects for ever."

"ROYAL INSTITUTION
  December 26th, 1839."

INDEX.

Absolute Zero of Temperature, 96
Air Engine, Joule's, 138
— Rarefaction and Condensation, 78
Albert Medal presented to J. P. Joule, 166
Amalgams, Investigations on, 138
Annals of Electricity, 33, 44
Atmospheres, whirling, Joule's Hypothesis, 95
Atomic volume, Researches on, 102
Atoms; necessity of motion, 14

Binney, E. W., portrait presented to the Manchester Literary and Philosophical Society, 166; death of, 168
Black's discoveries on heat, 18, 174
Boyle, Heat a kind of motion, 86, 174
Boyle's law, 82, 140; Explanation of, 87
British Association Meetings, 53, 67, 71, 107, 114, 135, 138, 141, 151—2, 160, 103, 109

Caloric, invention of, 18, 177, 184; hypothesis of, established, 91. See also Heat
"Calorific Effects of Magneto-Electricity, &c.," 67
Cambridge Philosophical Society, 116, 143
Carnot's theorem on the conversion of heat, 20; theory of the Steam Engine, 90; his law, 110; Sir W. Thomson on, 119; Rankine's equations in connection with his theory, 125
Carnot's function, form of, suggested by Joule, 128
Chemical affinity, intensity of, dependent on the gaseous or non-gaseous condition, 51; defined, 72
Chemical equivalents of electric action and heat, 47, 58
Chemistry, Extracts from Lavoisier's "Traité Elémentaire de Chèmie," 185
Clapeyron, E., on the Steam Engine, 88

Clausius, R. J. E., on the Mechanical Action of Heat, 128. Theory of Gases, 143, 152
Cold, action on Iron and Steel, 157
Colding's suggestions on the conversion of heat, 21
Collier, J., Portrait of J. P. Joule, 168
Condensation of air experiments, 82
Conservation of Energy, 2 *et seq.*
Copley medal, 156
Crompton, Dr. Samuel, 161

Dalton, Dr. John, 27; Joule's association with, 27; influence on Joule's work, 28; experiments on air compression and rarefaction, 80
Dancer, J. B., 81
Davies, John, teaches chemistry to Joule, 30; on animal heat, 71
Davy, Sir H., mechanical hypothesis of heat, 86, 124
Dynamo, invention of, 22

Earth, living force of the, 10; effect of falling into the sun, 11
Edinburgh, Royal Society of, 119, 123, 130
Electric action, Ohm's measure of, 24; mechanical equivalent of, 40, 47; heat equivalent of, 48; chemical equivalent of, 49
Electric engine, Joule's, 35
Electric measurement, absolute unit system, 41
Electric origin of heat, 49
Electricity, discoveries, 22; extract from Faraday's "Experimental Researches," 189
Electro-magnetic engine, 22; duty per grain of zinc consumed, 100
Electro-magnetic attraction, law of, 37
Elias, Dr. Daniel, attends Joule in illness, 162
Energy, conservation of, 2 *et seq.*, application of the term, 17, 92; concentration of, 104; dissipation of, 123

Faraday, Prof. M., electrical dis-

coveries 22, 23, 42, 56, 76, 87, 88, 109; extract from his "Experimental Researches," 189.
Faraday, F. J., 189
Force, not destructible, 88; destructible and indestructible, Roget's and Faraday's views, 187. *See also* Living Force.
Friction, heat produced by, 71.

Galvanometer, Joule's improvement, 37.
Gases, dynamical theory of, 112; their kinetic constitution, 95.
Graham's researches on the properties of gases, 95.
Gravitation as a property of matter, 3.
Gravity, specific, researches on, 102.
Great Eastern steamship, launch of, 147.

Heat, Living Force, and Matter, Joule's Lecture, 2; living force convertible into heat, 9; equivalency to mechanical power, 10; not a substance, 13; absorbed during change from solid to liquid, 15; growth of ideas concerning, 18; conversion into mechanical effect, Seguin, Mayer, and Colding's suggestions, 21; equivalent of electric action, 47; in metallic conductors, 47; developed during electrolysis, 48; electric origin of, 49; equality of heats resulting from combustion and electrolysis, 51; discrepancies between, 53; generated or transferred, 60; means of destroying, 64; mechanical value, 65; conversion into power, 69; produced by friction, 71; temperature in relation to, 79; latent, 83; dynamical theory of, 86; indestructibility of, 91; of gases, 113; inconvertibility maintained by Sir Wm. Thomson, 117; Rankine on the Mechanical Action of Heat, 123; Clausius on the same, 128; historic sketch of researches on, 133; further electrical experiments, 151; Joule's last experiments, 163; development of the knowledge of, 173; Black's discoveries, 174; Hooke's vibratory theory, 177
Herepath's theory of Gases, 112
Hirn, Gustave-Adolphe, experiments on the friction of water, 152, 164
Holmfirth Reservoir disaster, 142
Hooke, Robert, Vibratory theory of heat and light, 177

Horse, duty per grain of food, 101

Inertia as a property of matter, 4
Iron, resistance to induction of magnetism, 36; effect of magnetism on, 102; action of cold on, 157

Jacobi, Prof., discoverer of internal magnetic resistance, 40
Johnson, W. H., Portrait of E. W. Binney, 166
Joule, B. St. J. B., diary of, 26; taste for music, 31; marriage, 32; describes the Holmfirth Reservoir disaster, 142; railway accident, 147; stoppage of his brother's experiments, 148, 150; illness of J. P. Joule, 161
Joule, James Prescott, discoverer of the law of conservation of energy, 2; Lecture on Matter, Living Force and Heat, 2; birth, 25; studies chemistry under Dalton, 27; influence of Dalton on his work, 28; visits the Lake District, 29; chemistry lessons from John Davies, 30; accident with firearms, 31; First Research, papers in "Sturgeon's Annals of Electricity," 33; improves electro-magnets, 34; measures work, 35; finds speed limited, 36; standard galvanometer and law of electro-magnetic attraction, 37—38; mechanical equivalents of electric and chemical actions, 40; absolute units, 41; economic possibilities of the electro-motor, 43; Second Research, 55; communication to the Royal Society, 55; motive, 46; "Heat evolved in metallic conductors," 47; heat equivalent of electric action, 47; heat developed during electrolysis, 49; heats of combustion and electrolysis, 50; heat dependent on the state of the elements, 51; first paper before the Society, elected member, 53; research with Dr. Scoresby, 55; heat evolved during electrolysis of water, heat to give oxygen the gaseous state, 55; heat evolved by mechanical condensation of oxygen gas, 56; mechanical, chemical, and thermal equivalents of electro action, 57; Third Research, 59; heat generated not transferred, 64; "Mechanical value of heat," first determination, 66; meeting of British Association at Cork, 67; mechanical equivalent by friction, 71; generalisation, 72; efforts to convince the scientific

INDEX.   195

world, 74; papers on "The intermittent character of the Voltaic current in certain cases of electrolysis," and on "The changes of temperature produced by the rarefaction and condensation of air," 78; skill in thermometry, 81; experiments on latent heat, 83; new theory of the Steam Engine, 88; his knowledge of mechanical philosophy, 94; hypothesis of whirling atmospheres, 95; discovers the absolute zero of temperature, 96; letter to "Philosophical Magazine," 97; definition of gaseous hypothesis, 98; visits Dr. Scoresby, 99; paper on "Mechanical Powers of electro-magnetism, steam and horses," 99; essay for the Institute of France, 101; research with Sir Lyon Playfair, 102; appointed Secretary to the Manchester Literary and Philosophical Society, 103; Lecture on "Matter, Living Force, and Heat," 104; explanation of shooting stars, 105; paper accepted by the Institute of France, 107; meets Sir William Thomson at Oxford, 108; marriage August, 1847, 109, 110; experiments on gases, 112; his conclusions accepted by Sir W. Thomson, 119; and by Rankine, 127; final determination of mechanical equivalent of heat, 131; historic sketch of researches on heat, 133; elected Fellow of the Royal Society, 136; investigates amalgams, 138; invents hot-air engine, 138; joint investigation on gases with Sir William Thomson, 139; elected a Vice-President of Manchester Literary and Philosophical Society, 141; death of his wife, 142; honours conferred on him, 143, 156, 166; invents process of welding metals by electric current, 144; researches on solids and liquids, 145; elected on the Council of the Royal Society, 147; railway accident, 147; experiments on condensation of steam, 149; elected President of Manchester Literary and Philosophical Society, 150; death of his younger brother, 151; portrait by Patten, 151; description of him by the author, 154; receives the Copley medal, 156; papers on action of cold on iron and steel, 157; displays his character, 158; elected President of British Association for 1873, 160; first failure of health, 160; last experiments on heat, 163; Royal pension, 166; publication of his collected writings, 166; last paper to the Manchester Literary and Philosophical Society, 168; portrait by Collier, 168; death, 169; memorial statue in Manchester, 171; tablet in Westminster Abbey, 171.

Laplace's Theory of the Velocity of Sound, 106.
Lavoisier invents the term "Caloric," 175; Extracts from his "Traité Elémentaire de Chèmie," 185.
Light, Hooke's Vibratory Theory, 177.
Literary and Philosophical Society. *See* Manchester.
Living Force, Matter and Heat, Joule's Lecture, 2; Law of Velocity, 5; transference of, 6; not annihilated by friction, 8; heat, its equivalent when apparently destroyed, 8.
Locomotive Engine, 21.

Magnetism, effect on iron and steel, 101.
Magneto-electrical Heat, Researches on, 61; Laws of, 63.
Magnets, Experiments on, 35, 36, 168; maximum power of, 43.
Majocci's opinion on Gravitation, 3.
Manchester *Courier*, May 5th, 1847, Joule's Lecture, 104.
Manchester Literary and Philosophical Society, 27, 53, 55, 59, 77, 136, 138, 141, 145, 150, 156, 159, 168, 170, 172.
Manchester Meeting of British Association, 53.
Manchester, Victoria University, 166.
Mann, Dr. Dixon, attends Joule in illness, 162.
Mark, Ald. John, 171.
Matter, Joule's Lecture on, 2.
Mayer, M., suggestions on the conversion of heat, 21; proof of convertibility of heat into work, 86; experiments on temperature of water, 133.
Mechanical foundation of Physical Science, 2; equivalent of electric effect, 40.
Metals, welding by electric current, 144
"Micrographia," Hooke's, Extracts from, 177
Momentum, Joule's use of the term, 94
Motion of atoms, 14; perpetual, no fundamental objection to, 38

INDEX.

Motor, Electrical, Invention of, 22

Oersted's electrical discoveries, 22
Ohm's electrical discoveries, 22, 23, 24
Oxygen, affinities of, 51.

Paddle, use of the, 97
Patten, G., Portrait of J. P. Joule, 151
Perpetual motion, 38
Philosophical Magazine, 46, 53, 67, 77, 78, 97, 99, 101, 102, 105, 106, 111, 124, 138
Philosophical Society, *See* Cambridge and Manchester
Philosophical Transactions, 45, 127, 131, 139, 146, 149, 163, 167
Physical Science, Mechanical foundation of, 2
Playfair, Rt. Hon. Sir Lyon, researches on atomic volume, 102

Railway accident, 147
Rankine, W. J. M., hypothesis of molecular Vortices, 95; on "The Mechanical Action of Heat," 123; "Manual of the Steam Engine," 152
Rarefaction, Experiments on, 84
Regnault's Memoirs on the Steam Engine, 116
Repulsion as a property of Matter, 4
Resistance, internal magnetic, 40
Roget's views on the destructibility of Force, 187
Roscoe, Sir Henry, Secretary of Manch. Lit. and Phil. Soc., 150; intimacy with J. P. Joule, 156
Roscoe, Lady, portrait of Joule, 171
Royal Society, 45, 50, 131, 136, 139, 141, 147, 148, 171
Rumford, Count, on frictional heat, 19; observations on heat, 132

Salford, birthplace of J. P. Joule, 25
Schunck, Dr., Secretary of Manch. Lit. and Phil. Soc., 150
Scientific Papers, Joule's, Publication of, 166
Scoresby, Rev. Dr. Wm., 55, 77, 99, 100
Seguin, H., suggestions on the conversion of heat, 21; experiments on Steam, 133
Solids, Thermo-dynamics of, 146
Sound, theoretical velocity of, 106
Speed, fundamental limit to, 39

Stars, Shooting, explanation of, 11, 105, 111
Steam condensation experiments, 149; Steam-engine as a contributor to the discovery of the mechanical origin of heat, 20; increase of power, 68; new theory, 88; duty of a Cornish engine per grain of coal consumed, 101; Regnault's memoirs, 116
Steel, action of cold on, 157
Sturgeon, Wm., electrical discoveries, 22, 24; electro-magnetic engine, 33; Joule's memoir of, 145
Sun, effect of the earth falling into, 11

Tait, P. G., 152
Taylor, Dr., of Todmorden, treats Joule for spinal weakness, 29
Temperature, changes produced by rarefaction and condensation of air, 78; absolute zero of, 96
Thermo-dynamics, development of the theory, 128; application to gases, 139
Thomson, Sir Wm., application of the term energy, 92; names Joule's generalization, 105; on the Oxford Meeting of the British Association, 1847, 108; on the absolute thermo-metric scale, 116; on Carnot's theory of the motive power of heat, 119; on Rankine, 129; paper on "Dynamical theory of heat," 130; investigation with Joule on gases, 139; experiments on fluids in motion, 144; on latent heat, 146; awarded the Copley medal, 156

Victoria University, 166

Water, Electrolysis of, 55
Watt Monument, opening, 151
Welding metals by electric current, 144
Williamson, Prof., address to British Association, 1873, 160
Wind, cause of, 12
Windermere, Lake, depth of, 31
Woods, Thomas, M.D., conduct of, 149
Work, application of the term, 17; the measure of mechanical potency, 19; Joule's first measurement of, 35; dynamical significance of, 93

www.ingramcontent.com/pod-product-compliance
Lightning Source LLC
Chambersburg PA
CBHW020947230426
43666CB00005B/206